ENVIRONMENTAL ENGINEERING

ENVIRONMENTAL ENGINEERING

A Chemical Engineering Discipline

Edited by

G. LINDNER

Research Institute of Swedish National Defence, Sundbyberg, Sweden

and

K. NYBERG

The Swedish Engineers' Press, Stockholm, Sweden

D. REIDEL PUBLISHING COMPANY

DORDRECHT-HOLLAND / BOSTON-U.S.A.

CHEMISTRY

Library of Congress Catalog Card Number 73–75764

ISBN 90 277 0347 7

Published by D. Reidel Publishing Company,
P.O. Box 17, Dordrecht, Holland

Sold and distributed in the U.S.A., Canada, and Mexico
by D. Reidel Publishing Company, Inc.
306 Dartmouth Street, Boston,
Mass. 02116, U.S.A.

Printed in The Netherlands by D. Reidel, Dordrecht

TABLE OF CONTENTS

FOREWORD

Chemistry and its products today play an important role in almost all industrial activities. Chemistry has captured our homes. We are supplied with new articles in an ever-increasing stream. New uses are being discovered. Old products disappear.

Continuing and fast expansion is expected for the chemical industry in its proper sense. The reason for this is, of course, that chemistry has created products which meet requirements that we consider urgent or which in different ways make work easier, and make us more efficient, thereby increasing our standard of living in a wide sense: in terms of money, more spare time, social security, better education and better public health services.

But a high standard of living also implies a good living environment. A lot of what has been done in praiseworthy aspiration of a better means of support and an improved standard of living has involved a wasting of non-renewable natural resources. The products themselves or their waste products may pose a threat to the objectives we are trying to attain.

In view of the role of chemistry in modern society it is a natural thing that in public discussions on the environment the chemical industry is very often associated with damage to the environment. In these discussions the positive side is very often overlooked when people are confronted with the deleterious effects of pollution. But the problems are great. They must be solved promptly, preferably in such a way that we may keep the advantages that production has given and still gives. All possibilities and methods must be tried in order to minimize the discharge from the chemical industry into the air and water: new processes, change of raw materials, improved process engineering, improved processing equipment and improved regulation, and control and supervision of the production processes.

Knowledge about their own fields should afford chemists and chemical engineers the possibility of helping to solve environmental problems in connection with other fields of activity. Waste-water treatment techniques and sludge handling may be substantially improved by chemical agents and chemical engineering. Lakes have to be restored, and oil spills must be combated. Inexpensive methods are required to recover useful products from wastes. Last but not least it is necessary for us to neutralize – without the attended environmental hazards – wastes that cannot be made useful.

The demand in environmental protection work on the knowledge and inventiveness of chemists and chemical engineers is unlimited.

VALFRID PAULSSON

General Director
National Swedish Protection Board

PREFACE

Some ten years ago the alarm-bells concerning pollution reached an irresistible noise level. People all over the world concluded that there were draw-backs to the welfare evolution as well as limits in human living – factors that must be paid attention to. Today, there are some of us – although we are fairly well informed about the horrors of pollution – who look upon the situation in a much more positive way than at the beginning of this modern period of environmental consciousness.

For, nor the pollution problem is a modern one, neither is concern about the environment. You can find, even in very early historical documents, a lot of evidence of the world's ever-lasting pollution problems: heavy metals in drinking water, smell of fecal pollutants, bacteria, of course, and even the troublesome smoke of ignition fires. Living always creates living discharge problems. In our modern society, however, we live more comfortably than ever and thus have a bigger pollution back-lash.

But what has been the major difference in the last ten years? First of all it is knowledge. Even if we betray ourselves by saying that we know about every single situation, our over-all knowledge is better. Being conscious of the problems may sometimes be terrifying but it is above all a potential motivation for us to do something. Even if we don't know how to solve every single problem, there is a much greater motivation for finding out.

Technical development is different. Here you can find one of the very best examples of the old saying: 'Need is the mother of invention'. You need only mention a few – and by far not the most important – technical developments in fighting pollution of the environment. In less than ten years billions of dollars have been invested in thousands of different techniques in environmental measures.

The concept of pollution is different and has its own very well-known history. The first, second and third purification steps started from the mechanical problem of leading off unwanted pollutants, then the desire to eliminate sanitary risks from bacteria and finally the need of recycling. Pollution is today a balanced material's cycle problem. Chemists and chemical engineers have the best possibilities and qualifications to understand and solve this basic problem. One example of a chemical approach: if we look upon the world with an over-all chemical formula, we will find the use of energy for the decomposition of neutral substances like water and sodium chloride into local problems of acids and bases, or alkali metals and halogens respectively. Neutralization seems to be a natural solution but poses the problem of transportation.

Furthermore, if you agree with the idea of local problems and transportation need, you will also be aware of the concept of ecological balance. Since nature itself uses

NATO AND THE CHALLENGES OF MODERN SOCIETY

GUNNAR RANDERS

NATO Assistant Secretary General, Brussels, Belgium

Abstract. NATO, the North Atlantic Treaty Organization, is a well-known mutual security organization on the military side. What is not so well known is that NATO is involved in many activities to fight the forces which are threatening the quality of life. In this article these activities are discussed.

Over the last years some perhaps surprising newspaper reports have appeared, date-lined NATO Headquarters, Brussels:
  – Britain Enters World Safety Car Race,
  – Mathematical Model of North Sea under Study,
  – Conference Announced on Reducing Hazards from Earthquakes,
  – NATO to Tackle Water Pollution.
You may wonder why NATO as an organization should be concerned with such matters. The widespread public view is that NATO is a group of military forces watching the borders of Europe and the Atlantic Ocean. This is a narrow misconception. The very first paragraph of the North Atlantic Treaty cites the determination of the signatories to "safeguard the freedom, common heritage, and civilization of their peoples" and to "unite their effort for the preservation of peace and security". NATO is actively a mutual security organization. It has a highly successful record of defending the Western World against external forces which might destroy our countries. It should not be surprising that equally positive measures should be taken against those forces which threaten, from within, the quality of our life and the well-being of our citizens.

These civilian aspects of NATO are not new, but they have been largely unsung – which is one reason why I write this. Although preparation for a common military resistance to possible attack is basic to NATO's purpose, it is equally the civilian aspects of its work which deserve recognition. Daily consultations between the member governments at NATO Headquarters have been an increasingly key factor in insuring peace in the Western world. These consultations, carried on on a continuing basis, deal not only with external threats but with all factors which make up the inter-relations of the member countries themselves. These relations are discussed, negotiated and adjusted across the entire spectrum of mutual problems. Economic and political relations, not only with each other but with the non-NATO world, are under continual review.

Cooperation in defense research, weapon systems development and the immense infrastructure of pipelines, roads, airfields, warning systems and the like is probably not surprising to you. But it may be news that there has been, for the last fourteen years, a vigorous program in NATO for the support of fundamental science. Funded at a current level of about $ 5 million a year, it stresses cooperation among scientists in the member countries with the goal of improving man's knowledge of himself and his world. Some 50000 individual scientists have participated in this program; the main product is more and better science. A side-product is the development of life-long friendships and understanding among an increasingly influential segment of our society. Similarly, economic analyses are carried out on a co-operative basis, and a modest program of cultural and historic studies, highlighting our common heritage, has been underway for several years.

This brief introduction to today's NATO has implicit in it several answers to the question of a NATO role in dealing with our environmental problems. First, by experience, NATO has impressive qualifications for accelerating the transfer and application of technology. Second, NATO can work fast, and is geared to action. Third, it is able to command the attention and response of governments. Underlying all of these, though, is my first and chief point. NATO is an expression of the most vital ties between Europe and North America, it is thus ideally suited to a common task which requires sensitivity, mutual understanding and a capacity for international teamwork.

It was this kind of thinking which caused President Nixon, in 1969, to propose a new effort in NATO to deal with the problems of modern society. His suggestions were well received, and by autumn of that year the North Atlantic Council established a new Senior Committee, the NATO Committee on the Challenges of Modern Society, known, for obvious reasons, as the CCMS.

Now, why are we, rather suddenly, devoting such attention, and international attention at that, to our physical and social environment? The most simple answer seems to be the quite recent size and scope of the problems. These are due to the exponential nature of the curves of our population growth and our technological growth. A corollary is the increase in urbanization through the world.

Let me just briefly indicate the dimensions of the situation. It took 1700 years, since the beginning of the Christian era, to double the world's population from about 250 million people to 500 million. It took only another 200 years to double it again, and then 60 years to redouble it, with a present level of over 3.5 billion. The next doubling is expected to take about 30 years. And these are geometric projections, i.e., in the next 30 years as many people will be added to the world's population as were added in the last 300 (while the rate varies very much from country to country, there is hardly a nation which will not be drastically affected by these changes in the very near future.)

At the same time, technological progress is increasing even faster. It took 112 years for the scientific bases of photography to be converted into social use. A century later, the time it took for radio to be put into application was reduced to 35 years. Television took 12 years, and the principle of the transistor took three years.

Perhaps the briefest way to document this point in a global sense is to look at energy demand. This is growing at the rate of 4% annually in the highly developed countries, thus doubling every 19 years, i.e., significantly faster even than population growth. Much of the energy is used in processing matter for man's use, and the remainder is employed in direct services of transportation, communication and the like. It is the dispersion by this ever-increasing population of its ever-increasing matter and energy-production by products that is causing our environmental problems.

The nature of such effects on our environment is known to all of you, but again a few numerical indicators may not be out of place.

– At its source in Switzerland, the Rhine River has some 30–100 harmful bacteria per milliliter. At Bodensee (Lake Constance) the figure is 2000; at Bonn it is 24000 and when the Rhine enters Holland the figure is 100000.

– Ice cores from Greenland show 220 nanograms/kg of lead in ice deposits in 1965, 80 nanograms twelve years earlier and 20 nanograms two hundred years before that. That this phenomenon is due to man's activities is shown by comparison with samples from the Antarctic which is largely untouched by the winds and the currents which sweep over the populated Northern Hemisphere. There the lead concentration has stayed at a constant, low level for over 25 centuries.

– The number of sea trout captured per acre off the Texas coast declined regularly from 30 to 0.2 over a six-year period, and this has been correlated directly with DDT concentration in the waters.

– The first oil well went into production just over 100 years ago, but until 1900 oil production was measured in hundreds of tons. Today it is estimated that we deliberately dump between three and ten millions tons into the oceans each year.

– Last year in the U.S.A. traffic fatalities exceeded 1000 a week, injuries 10000 a day, and economic losses $ 1 billion a month. European countries are no better off. Counting deaths per 100000000 vehicle miles, the U.K. fatality rate was 13% higher, Germany and France were almost 250% higher, and the Netherlands was 300% higher.

Similar data on everything from noise pollution to the number of inmates of mental hospitals to deaths from fire to solid waste accumulations which threaten to bury us can be found in any newspaper or magazine.

What can man do about these problems which seem to be so large as to be overwhelming and insoluble? Unfortunately there is a lot we don't know. We don't know the real fate of oil in the oceans – the degree to which it is degraded by sunlight and microorganisms or the degree to which it causes irreversible damage to marine life. We don't know the real role of increased carbon dioxide accumulation in the atmosphere – whether it will lead to the ultimate warming up or cooling down of our planet, with catastrophic effects either way on the polar ice caps and the level of the seas. We don't know the effect on the food chain of. e.g., the 496 different synthetic organic chemicals recently found in fresh water samples in the U.S. We don't know the nature of the complex inter-relationships between point source pollution, dispersion mechanisms and ultimate effects on man or his planet.

But on the other hand, there is much that we *do* know.

And this is what the NATO CCMS is all about.

There are two major attributes of this new NATO Committee which give it its unique character. First, it is geared to *action*. The output of the Committee's work is recommendations to the member governments of the Alliance on specific actions which they may take, in the field of legislation, public education, executive programs and the like. The CCMS does not denigrate the importance of research by any means, but such research is already being carried out in the public and private sector, on a national and international basis. In fact, within NATO itself the aforementioned Science Programme is admirably structured to support cooperative endeavours aimed at the production of new knowledge. The foremost function of the CCMS is to turn existing knowledge into practical action within a reasonable time.

A second characteristic is the Pilot Country concept. Here one nation does the substantive work on a given subject. While other countries may cooperate as co-pilots, or just to the extent of making their own data available, it is the Pilot Country which bears full responsibility for the carrying out of a project. This concept has two distinct advantages. It avoids an intolerable load on the NATO Secretariat, which has neither the expertise nor the manpower to do the job, and which has no wish to compete with national authorities in the provision of this expertise which is already in dangerously short supply. Further, it insures that the work is properly motivated. A country volunteers as a Pilot only if it has serious intentions to do the work well, and is willing to commit the necessary human and financial resources. I can assure you that this is far more effective than the cooperation sometimes grudgingly offered just because a nation is a member of a given organization.

A third characteristic of the CCMS is worth noting because it is somewhat unusual in the NATO context. This is the only NATO body which was established on the fiat that all of its work would be unclassified and freely available to all. The pilot projects themselves are carried out openly, sometimes with the participation of non-NATO nations, and the completed studies and recommendations are widely distributed and publicised.

The CCMS, then is a major Committee of NATO. At present it meets in plenary sessions twice a year, with delegations from capitals including Ministers, Ambassadors, senior policy officials and technical experts. As necessary between the plenaries, it may meet at the level of representatives from the fifteen permanent national delegations housed in NATO Headquarters. The job of the CCMS is to consider and approve proposals for Pilot Projects, to review the progress being made on these projects, and to consider and endorse those interim or final recommendations of the pilot projects which are deemed significant, practical and acceptable. Recommendations approved by the CCMS then go to the North Atlantic Council for final vetting. This double-barrelled approval process is significant because it ensures that recommendations which are finally passed to the member governments for implementation have been seriously considered at both the technical and political level, and that they are backed by the full authority of the Council which enjoys considerable influence in all the NATO capitals.

1. Pilot Studies

There are currently active pilot studies in seven areas: air pollution, coastal water pollution, inland water pollution, road safety, environmental and regional planning, advanced health care, and waste water treatment. Let us look at each of these.

1.1. AIR POLLUTION

The basic element in this project involves the sampling of air over Saint Louis, Missouri; Frankfurt, Germany; and Ankara, Turkey. These three cities represent a variety of air pollution problems. Saint Louis embodies the characteristics of a regional metropolitan centre on the borders of Missouri and Illinois. Air pollution resulting especially from industrial and automobile emissions can be tested in a regional setting. Frankfurt represents an air pollution problem created by heavy industry in a Ruhr-based metropolis. Ankara provides a model of a large urban centre located in a depression, where coal with a high lignite content is burned. This results in a very heavy pall of smoke hanging over the city.

The first phase of this study involved development of air quality criteria relating to sulphur oxides and particulates. Even more significant was the development of a methodology which might be applicable in testing air quality over any metropolitan area. As a result of the completion of this first phase work, the North Atlantic Council gave its approval on 20 January, 1972, to an air pollution resolution which resolved that member nations will:

(1) Take cognizance of the need for dynamic efforts toward enhancing and maintaining the quality of the air:

(2) Endeavour to use, where appropriate, for setting up national air quality management programmes, the systems methodology generated by pilot study; and

(3) Promote co-operation among members when air pollution problems are common to national boundaries.

The first phase of this pilot project was so successful that the World Health Organization is using some of the results. Today, four additional NATO countries – the Netherlands, Norway, Italy and France – are participating in the second phase work involving especially nitrous oxides and carbon monoxide (auto-emissions). This final phase of the pilot study is expected to be completed this year, at which time a final report will be issued.

1.2. COASTAL WATER POLLUTION

This pilot study consists of three elements: ocean oil spills, a mathematical model of the North Sea, and ocean dumping.

1.2.1. *Ocean Oil Spills*

An Ocean Oil Spills Conference was held under the aegis of CCMS and sponsored by Belgium in November 1970. As a result of the resolutions and recommendations emerging from that Conference, which was attended by over 100 representatives from 14 member-countries, NATO nations committed themselves "to achieve by 1975 if possible, but not later than the end of the decade, the elimination of intentional discharges of oil and oily wastes into the sea". They also resolved to seek early implementation of the 1969 amendments to the IMCO Convention for the Prevention of Pollution of the Sea by Oil and "to work urgently on programmes to minimize the risk and consequences of accidental spills".

This resolution, as well as three others dealing with scientific research, technological counter-measures, and mutual assistance in countering oil-spill damage, depended upon action at two levels. First, NATO nations pledged themselves to take more vigorous national action through legislation, research, and contingency planning to assure that all possible measures would be taken to prevent or to minimize the harm done by ocean oil spillage. Secondly, that steps be taken at the international level, especially through the United Nations and its specialized body IMCO, whereby NATO nations would press for speedier and more effective action. Progress in implementation of the ocean oil spills resolutions and recommendations is being monitored by Belgium on behalf of CCMS. Twelve members of the Alliance have submitted reports since September 1971 which show a considerable degree of action to implement them, both on the national and international levels, especially through IMCO. There will be a continuing follow-up annually which will serve both to remind NATO nations of their obligations under the oil spills resolution as well as provide encouragement for continued attention in NATO capitals to this urgent problem.

1.2.2. *Mathematical Model of North Sea Pollution*

Under Belgian leadership, an extensive oceanographic research programme is under way in the North Sea to develop a mathematical model of its pollution. This is designed not only to identify the current character and extent of areas of pollution, but also to predict how that pollution will develop and where it will be located in future. Accordingly, if this mathematical model is sufficiently accurate, it could provide a key to the future nature and location of pollution areas in the North Sea. Finally, such a mathematical model might be of considerable value in providing a methodology to attack pollution problems in other bodies of water such as the Mediterranean, the Baltic, and the Adriatic seas.

1.2.3. *Ocean Dumping*

NATO's oil spills resolution is but a reminder that oil is only one of the many contaminants discharged into the seas, and that consideration must also be given to other forms of ocean dumping. The latter has been the subject of discussion in the context of the United Nations' Conference on the Human Environment and has resulted in a number of regional agreements. However, Belgium and the CCMS are maintaining a careful watch over this subject, and if required, NATO might become more actively engaged. Meanwhile, the catalytic effect of CCMS's action and its follow-up helps focus attention on the continuing need to combat oil spillage into the world's oceans.

1.3. INLAND WATER POLLUTION

Canada and the United States have undertaken to provide as the principal element in this pilot study a comprehensive river basin planning and management programme for the St. John River, which flows between New Brunswick and Maine. The areas to be covered include comprehensive water quality planning; administrative arrangements for planning, implementation, and follow-up and evaluation; financial arrangements for planning and implementation; public involvement and information exchange for planning and implementation; and international aspects. When this study is completed in 1974, it is expected to serve as a model applicable to other international river basins. This model will be sufficiently broad to include political, economic, sociological and environmental factors as these enter into planning with respect to inland water pollution and river basin development in an international setting.

1.4. ROAD SAFETY

Consisting of seven projects, this pilot study is one of the most active in NATO/ CCMS. The United States which directs the overall work, is, in addition, the project leader for the Experimental Safety Vehicle (ESV). Bilateral technical agreements relating to ESV development have been entered into between the United States and six countries, four from NATO and two non-NATO. These include the Federal Republic of Germany, Italy, the United Kingdom, France, Japan and Sweden. The objective is to develop ESV prototypes in both the 2000 lb and 4000 lb categories with a view, among other things, to incorporating safer design features in production-model automobiles. A number of prototypes were put on public display for the first time at the Third International ESV Conference held in Washington in June 1972. The next ESV Conference is scheduled to be held this year in Tokyo. This pilot project is a prime example of how CCMS has played a catalytic role in speeding the collaborative development of ESV prototypes. The CCMS pilot project provided an organizational and administrative framework to enhance multi-national co-operation in developing ESVs. This has been far more efficient and less costly than a series of bilateral co-operative groups would have been. More importantly, CCMS has helped speed the early incorporation of safety features from ESVs into production-model cars.

Belgium, as director of the project on *Pedestrian Safety*, is concerned especially with improving the collection of data and its comparability with a view to assisting national authorities in improving relevant safety measures. *Canada* leads the project on *Alcohol and Driving* which is focused largely on the planning of roadside surveys in several NATO countries using the breath-test to determine the alcohol in a driver's breath. This technique, rather than the older blood tests, might provide a more feasible means of giving roadside tests to check suspected offenders. *France* is leading the project on *Road Hazards* which is designed to amass data from NATO countries which can be used in preparing a future report to ameliorate the problem created by road hazards on many heavily travelled highways. The *Federal Republic of Germany* leads the *Motor Vehicle Inspection* project which has organized workshops to demonstrate the system in use in the FRG, to see whether it is applicable in other countries and to promote an exchange of information on systems operated elsewhere. *Italy* leads the project in *Emergency Medical Response* where a variety of national data and a number of suggestions from NATO countries will serve as the basis for a future symposium and report which might lead to improved medical response to highway accidents. The *Netherlands* lead the project on *Accident Investigation* which includes as a major element establishment of a standard form for the use of accident investigation teams. Fifty such teams in six NATO countries have been collecting data on road accidents which will be evaluated and then processed by computer to serve as the basis for a future report.

As work proceeds in all seven of the Road Safety project areas the findings and interim conclusions are contributing to national road safety programmes. Increased impetus is given to taking practical steps on the basis of preliminary but effective work. The entire pilot study is expected to terminate its activities during 1973 and produce a final report and recommendations in early 1974.

1.5. ENVIRONMENTAL AND REGIONAL PLANNING

Whereas other CCMS activities might be classified as 'sector' or 'vertical' approaches to environmental problems, this pilot study is concerned with the interconnection between environmental problems in the framework of regional development.

The study accordingly aims to examine the relative efficiency of regional and local institutions in dealing with environmental policy, and the relationship between environment and territorial, economic and social development. As a co-pilot, the U.K. is studying the role of various levels of government in regional development. Other countries, especially the US, are also contributing to the data base being developed by the pilot country. A report has been produced in November 1972 which might offer initial guidelines as a model in the planning of regional development taking into account more fully broad environmental considerations.

1.6. ADVANCED HEALTH CARE

This pilot study was organized and approved by the CCMS in plenary session in November 1971. The project was informally discussed with WHO officials in Geneva who sent an observer (as did the OECD) to the second expert meeting of this project in Brussels, in February 1972. This pilot study comprises five projects as follows:

1.6.1. *Automated Clinical Laboratories*

Led by the United Kingdom, this project aims at helping NATO nations to cope better with the growing burden, through increased automation, of laboratory tests presently requiring large inputs of highly-skilled, scarce and costly manpower.

1.6.2. *Organized Ambulatory Health Services*

Headed by the Federal Republic of Germany, this project proposes to examine means whereby large groups of patients who are not hospitalized can be classified quickly according to their particular needs for medical attention, thereby facilitating referral for diagnosis and treatment.

1.6.3. *Emergency Medical Services*

Italy leads this project, which is designed to ensure more efficient use of scarce and expensive facilities and specialized manpower available at hospitals not only to give emergency medical care in road accidents, but will embrace a wider field, including home industrial and other accidents.

1.6.4. *Medical Surveillance Methodology*

This project headed by Canada, aims at seeking ways in which physicians might improve their efficiency on the basis of surveillance and evaluation by their peers. An essential factor involved is that eighty percent of all health costs are generated as a result of physicians' actions and determination.

1.6.5. *Drug Rehabilitation*

Led by the United States, this project seeks to enlist NATO co-operation in devising techniques to combat drug addiction and to aid in rehabilitation of those who have become addicted.

1.7. WASTE WATER TREATMENT

This pilot study which was approved by the CCMS in plenary session in April 1972, comprises two principal elements:

(1) The U.K. has undertaken to build a pilot plant near Birmingham which will employ the physical-chemical process in the purification of waste water (sewage).

(2) The Federal Republic of Germany has undertaken to test out the oxygenation process in the purification of waste water (sewage) at a plant in the Federal Republic. France is expected to co-operate closely with the FRG in developing the oxygenation process, as well as embarking upon an independent line of approach. The U.S. is collaborating with the participating countries in developing and evaluating the most advanced technology. It is hoped that as a result of this pilot study, NATO nations will have available to them a number of optional technologies from which to choose in meeting their waste water problems.

2. Conclusions

As a two and a half year experiment, CCMS has already demonstrated its capacity to encourage more prompt action in dealing with urgent problems of modern society. Particularly in the areas of ocean oil spills, air pollution, road safety and disaster assistance, CCMS has fostered earlier national and international action. This catalytic role, of course, will require continuing follow-up to assure effective implementation.

Thus, two series of steps – at the international and national levels – provide the blueprint for an effective NATO role coping with environmental and other challenges of modern society. Relying upon the political cohesion and strength of the Alliance and the effectiveness of commitments undertaken within NATO, compliance is encouraged through the process of consensus and persuasion which is inherent in NATO co-operation and collaboration. Although there are no sanctions, through

NATO's political process and the agreement to report progress in implementing the terms of CCMS resolutions, effective action is encouraged.

There are also ancillary effects of CCMS's work. Obviously, it was inevitable that NATO nations would give increased attention to their domestic environmental problems and the organization of their central governmental apparatus to deal with them. However, under the impetus of the establishment of CCMS and especially its twice a year plenary sessions, some NATO nations have been encouraged to speed up their attack on domestic environmental issues. The CCMS plenary sessions provide a valuable opportunity for an exchange of views on how NATO nations are handling their national environmental problems. It also provides a valuable forum for frank consultation on issues of broader international concern. For example, at the April 1972 CCMS plenary session, two subjects on which there was an animated and valuable exchange of views were the UN Conference on the Human Environment and the global problem of ocean dumping as related to that Conference. Accordingly, we can add a new element to the consultation process in NATO, namely consultation on environmental and related problems.

To sum up, the Committee on the Challenges of Modern Society has demonstrated its capacity to deal in a pragmatic manner with real and urgent problems affecting modern society. CCMS provides a flexible, open ended, and completely voluntary means whereby those NATO nations interested can participate in collaborative efforts to solve problems of mutual concern. The success of this effort will depend upon the willingness and resources which these nations bring to the solution of these problems.

ENVIRONMENTAL ENGINEERING – A SOURCE
OF NEW INDUSTRIES

C. P. HICKS

Manager, Pollution Control Division, Alfa – Laval AB, Tumba, Sweden

Abstract. The role of the chemical engineer in environmental engineering is outlined followed by a discussion of industrial activities with its background of economy.

When I sat down to think about this I realised that it was the chemical engineer to whom this was a challenge and not the capitalist as one would suppose.

I need hardly remind you that the reason we are talking about environmental engineering is because pollution, together with other factors, is changing or upsetting the ecology of the world. However, we must remember that it is mankind that is upsetting or changing the balance. In animal terms he was not expected to have intelligence, it was ecologically logical that pestilence and disease would keep his numbers under control. The fact that he has intelligence has caused him like the lemmings and locusts to multiply too fast. This intelligence has halted disease and so far prevented him from starving. He has done this by producing his food, water and material needs but in doing so he has changed the face of the world.

To carry through these changes man has organised himself like the bee hive into groups of specialists, the farmer, the politician, the doctor etc. One specialist group who has had to serve all these people in the primary and secondary sense is the engineer. He has devised the tractor, the motor-car and all other machinery of to-day's existence. Two things have happened as a result. The first is that the engineer and particularly the chemical engineer has done what the community asked him without receiving thanks. The community asked him to produce bulk fertiliser which grew the wheat, not knowing this would destroy the lake and so on. Engineers appreciate the delicate balance of ecology when they understand it, for it is a balance which is like the balance of forces in engineering just as the metabolic system is like the chemical reaction.

It is the engineer who has prevented the community from destroying itself with its own excreta and produced systems which produces uncontaminated water.

The engineer, by his upbringing appreciates his environment. Bridges are beautiful, sometimes electric cable pylons are not... but the engineer would gladly put them underground – if the community would pay. Engineers can silence aircraft engines – but will the community pay? Engineers can do most of the things that the community demands if it is prepared to pay. Sometimes engineers do things which cause unforseen results, but this is through ignorance of the effect rather than through intention.

I find that far too often the chemical engineer is blamed for doing what was demanded of him by the community.

G. Lindner and K. Nyberg (eds.), Environmental Engineering, 13–17. All Rights Reserved

The problem is more complex that it is made to appear by the journalists. It brings us back again to the question of ecology. We all know what the first steps are. We must stop pouring SO_2 on the streets and into the air in a wide variety of ways. We know the effect of oxygen deficiency in the lakes and we are beginning to realise the effect of phosphates where they should not be. But we must realise that this is not a stable world, we never will know what is the norm for life. Is the human being the intruder, or is bacteria? The world is changing just as much as we are changing the world. The dinosaurus might have said that it was being destroyed because the world was being polluted by these animals invading a world built for reptiles. If an alien came to this world it might assume that it was wheat that ruled the world. It is cared for and nurtured and land is torn up for it to live. It is cherished and fed like nothing else, so surely it must be the most important thing on earth. We can take steps to stop the immediate problems, we can take steps to prevent further problems in the short term but can we deal with the situation in the future? I am not sure that we can, the sociologists and the bureaucrats cannot control the population and its demand so we chemical engineers will have greater demands thrust upon us, but there will not be enough of us for the future to meet these demands. You cannot easily make an engineer out of a botanist, but I believe that if you can suffer the requirements to become a chemical engineer you should be able to understand the requirements of the environment.

The challenge to the engineer of to-day is first to explain to the community what he does, how he thinks and for the world to understand that he is not the destroyer of their environment but the one saviour that is available to mankind. I realise many of you are thinking already... this is not necessary, everybody knows what we do and why we are what we are... but gentlemen, you are wrong. We are an unknown profession, and furthermore a profession totally unappreciated... rarely in the pages and pages that have been written in the popular press about pollution has it been appreciated what we are and what we have done to keep the world as clean as it is. It is obvious that we have failed because when you look at the universities and technical schools of the Western world, you find it is not full of chemical and other engineers and they are not getting the quality of students they should... School children and teachers alike in the lower schools do not see it as an exiting career, challenging and worthwhile. The youth of to-day with ample opportunity often wants to do some good in the world, but they would laugh at you if you suggested that they could fulfil this ambition by becoming a chemical engineer.

I believe we must publicise our activities in the right way. Tell the public what we have done for them and thus ensure our professional position. We must show them that our job has been producing the goods, and food for them. That given the money we can meet the needs for which community has asked. We have been asked for many things in the past and our predecessors have given them to the community... clean water, steel, motor cars, TV, cheap fibres, the removal of manual labour, and now we are being asked to control the overall activities of mankind so that we can stabilise a changing world.

We can do this but we have to adjust our thinking to withstand outside pressures. We have to consider our position very carefully. Pollution control cannot be done by legislation alone, it has to be done by our professional bodies accepting that this is part of our ethical code. Just as safety is accepted so must pollution be. We refuse to build an unsafe process unit so in the same way we must refuse to build a plant which would cause pollution. This is not going to be easy. We are beginning to realise when we question these standards what our predecessors had to do when they started to fight for safety precautions. We engineers have been reluctant to fight and speak up for this approach. There is justification for Ralph Nader's request that "engineers should end their silence in this matter".

You may wonder what this has to do with a source for new industries, Well, it has a lot. To start a new business one has to have certain conditions. Among them are:

(a) A Market demand,
(b) Capital,
(c) Resources to satisfy the need or demand.

The demand is to prevent pollution and humanity has asked us to do this. This is clear.

The resources to a great extent are going to be professional engineers and their position which I have just discussed. The capital must be found and can be.

The community, each one of us will have to pay either in higher prices or taxes. As far as one can tell this has been accepted... but the real worry is what will everyone get for their money? New industries if we are wise.

It seems that there are five different types of new industry that can be created.

(1) Extensions of existing industries,
(2) Special waste consuming industries,
(3) Environmental engineering equipment manufacture,
(4) Specialised contractors,
(5) Environmental engineering consultants.

1. Extensions of Existing Industries

This as such is not a new industry but the completion of a process. It will, however, create new opportunities. Opportunities for pollution control engineers. It will also create new parts of an industry, the recovery of stack SO_2 is already well under way, but there are many other ways in which by-product industries might grow. Factories might have farms beside them so the effluent can be disposed of by irrigation. Households can be heated by low grade heat from the adjacent plant. As such operations are simply the completion of the process the industry itself should finance it itself wherever possible.

2. Special Waste Consuming Industries

Waste is far too often wasted. Here is a chance for the inventor. How to use old glass bottles... grind them and turn them into non slip floors perhaps. What to do with community sewage waste solids... ideal for specialised fermentation perhaps. The whole field of fermentation is ready to exploit many wastes.

Already to-day there are many waste consuming industries, but generally they are looked down upon socially and technically.

The recovery of metal scrap is one. Another is the reuse of motor car oils. This is a rather simple refining process... processed used lubricating oil was exported to the middle East for many years. Big bus operating companies used to centrifuge their oil before reuse, but such activities were discouraged by the oil suppliers for obvious reasons, but they did solve a disposal problem.

Because these industries are looked upon as inferior, they tend to be 'poor' and do not encourage scientific operation.

Here is a real chance for community money to be well spent. Such industries tend to be risky capital investment and so are starved of capital. A government development corporation should be formed to give such industries help in the form of equity financing and in the form of government contracts by guaranteeing some of the initial sales.

3. Environmental Engineering Equipment Manufacturers

The comments I now make can be misinterpreted as a plea for my own company... but this is not so. I just speak from bitter experience! The suppliers to this type of industry have a terrible time. We are often criticised for not producing the highly specialised machines needed.

The development costs of these specialised machines often amount to between 30% and 50% of the total projected sales. It is an uncertain market, for such specialised machines can usually be copied rather easily. Here is a real case for a direct grant. The governmental development corporation should draw up specifications for such items of equipment, put them out to tender to suitable firms or individuals as development projects and then pay for such development work. The spin off from such a method is substantial because such schemes help to sustain the base loads for large companies research organisations and will also encourage inventors and innovators to exploit new approaches on their own.

The opportunities are many, cheap smokeless incinerators, simple scrubber systems, etc.

4. Specialised Contractors

These come in all shapes and sizes, they are all service industries, the collection of scrap metal at one end and the construction of complete plants at the other. In both cases it is usually working capital that they are short of and here again I would prefer to see government participation.

There is one job that I wonder if it is right for the state to be so closely connected to. This is the collection of household and industrial garbage. When run by the community it tends to be a rather negative operation with the result that it is usually a cost burden. For example, there is no encouragement for the citizen to separate his paper and metal waste, now if this collection was in the hands of private operators it would be possible for them to pay for segregated scrap, thus benefiting everybody.

5. Environmental Engineering Consultants

We are going to need these people NOW and I do not see how we are going to operate all these schemes needed without the formation of more consultants for this work.

– They must be licensed by the authority to give consultative advice.

– They must be able to give qualified advice and recommendations especially to the smaller industrial operations, and finally:

– They should approve pollution control systems.

Advice to the small industry is very important. In England just after the war, fuel was in very short supply and an organisation was formed called the National Industrial Fuel Efficiency Service. It gave a survey of a plant free, then if it was asked to control the operation it charged the client. I believe that we could develop this same idea by using environmental consulting engineers... payment for the surveys being by the local authority.

This is not sufficient encouragement for Chemical Engineers to start up in this type of operation and I suggest that the further encouragement should be given in the form of tax relief to chemical engineer consultants who wish to start up in the business.

To summarise the situation... the resources can be found, the need is there, the capital or cash is available BUT the money must be used efficiently to help create these new industries and this needs urgent planning otherwise we will see money being wasted – and that perhaps is the very worst waste problem as that cannot be recovered or reused.

WHAT ABOUT WASTES AND HOW TO REUSE THEM

ANDERS S. H. RASMUSSON

Royal Institute of Technology, Stockholm, Sweden

Abstract. The sources of different wastes are classified together with methods to take care of them. The problems are illustrated from the sulphur field with possibilities of reuse of different compounds.

I would like to remind you of two alarming events from 1971. July 15 the ship 'Poona' got on fire in the harbour of Gothenburg, Sweden. The fire was probably caused by rapeoil and sodium chlorate loaded together in the ship in a dangerous way. Several explosions happened and the fire was hard to extinguish. Afterwards the remains of the cargo were cleared away. Originally more than 700 drums of arsenic wood preservatives were included. After about one month it was discovered that 60 of these drums had been transported to the city dump situated in the neighbourhood of the city drinking-water reservoirs meaning an immediate danger to these reservoirs. Another 26 drums were transported to a private refuse dump before the authorities could act to prevent further dumping.

August 10 German papers unveiled another chemical waste scandal. Cyanide wastes from all Europe, including Sweden, Finland and Italy, had been collected in Bochum city dump in the heavy populated Ruhr district of W. Germany. It was established that more than 15000 drums had been put there e.g. by a firm offering destruction of chemical wastes. The cyanides were of immediate danger to water resources. 350 soldiers, policemen etc. have worked on the dump, collecting and destructing the wastes. A 10 km pipe-line has been built to carry off the drainage water from the dump. This event was not an exceptional one in W. Germany. The same week another five occasions were published in German papers.

1. Wastes and the Natural Ecosystem

When talking about wastes we often focus our attention on such wastes as car-wrecks, glass bottles, paper bags, plastic packages and cans. As shown in Table I, chemical wastes are a small part of all wastes but they could be much more dangerous than other wastes through their influence on drinking water, vegetation and animals or other parts of the biological system. Table II gives some actual figures of the amount of more essential chemical wastes in Sweden just published by a government committee investigating the problems of chemical wastes and their treatment.

Some decades ago we could still believe in the ability of nature to take care of dangerous wastes when these were handled with a minimum of care. The discoveries of a DDT-content in animals all over the world, the effect of biocides on birds, the circulation of heavy metals in nature including man and especially the Heyerdahl report on the heavy oil pollution of the Atlantic have however demonstrated to us

G. Lindner and K. Nyberg (eds.), Environmental Engineering, 19–31. All Rights Reserved
Copyright © 1973 by D. Reidel Publishing Company, Dordrecht-Holland

TABLE I
Solid wastes U.S. 1970/'71

kg/year pro capita 16000 total wastes	89%	Mine tailings, smelter slags, dredging spoil, agricultural wastes
	6%	Industrial wastes
	5%	Municipal wastes
960 industrial wastes	50%	Chemical industry
	50%	Other industries
800 municipal wastes	50%	Paper
	5%	Plastics
	45%	Others i.e. glass, rubber tires, metal cans.

TABLE II
Chemical wastes Sweden 1970/'71

	Kg/year pro capita
Petroleum wastes	25–35
Oil contaminated soil	10–60
Sludges from gas stations	1–6
Solvents	1–6
Acids and alkalies contaminated with heavy metals	3–15
Other acids and alkalies	6–25
Solutions from plating industry	1–6

beyond doubt that this ability of nature is limited. The living standard of man is rising, so is the production of wastes pro person. Without counter-measures the natural resources could soon be exhausted. This has been illustrated by Malaska in Figure 1. All the time we take natural resources out of the ecosystem converting them to wastes which are put back into another part of the ecosystem. With growing population and growing needs pro person the unconsumed part of the ecosystem will diminish with time until the biological nature is disturbed and at last destroyed. The timeslope of this curve must be reduced. Of course we cannot have a biological life without destroying natural resources but this destruction must be diminished by all means. The path to an acceptable solution is illustrated in Figure 2. A technology for reusal of wastes will diminish the natural resources taken out of the ecosystem as well as the wastes put back into that system. The principle is not a new one; it has been adopted for a long time by all scrap-iron dealers, rag-and-bone dealers, etc. To-day, however, we have to expand the use of this principle to all kinds of wastes.

2. The Origin of Wastes

The human activity can schematically be divided into two types of processes here called production processes and consumers processes. The waste treatment is to be

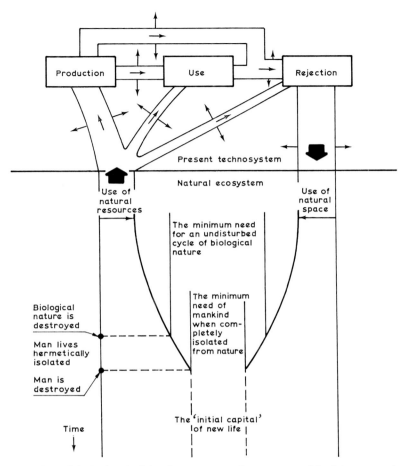

Fig. 1.   Illustration of the basic principle of our present technosystem and the future scenario upon it.

considered as a kind of production process. In both types of processes wastes of all kinds are produced. In consumers process (Figure 3) products are converted or used. Wastes produced are moved away from the customer by means of air, water or mechanical devices, eventually treated before dumping or reuse or dumped without any treatment. The origins of wastes are collected in Table III. Some wastes are formed during the biological or technical processes involved, e.g., excrements and detergents, other wastes like car-wrecks and scrap-iron depend on the limited life-time of products, still others on the high living standard and its high demands for papers, packages and other short-lived products, some depend on misuse of products.

The production process (Figure 3) produces wastes and products but the principal types of wastes are the same as in the consumers process even if the components could be quite different. We often talk about air pollution, water pollution and

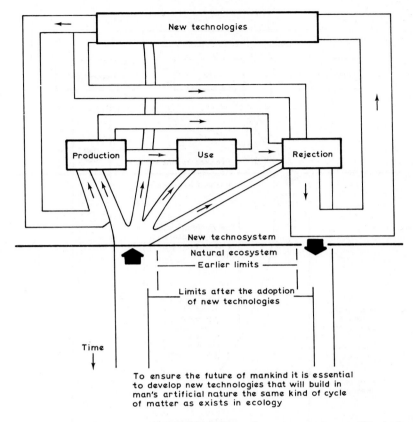

Fig. 2.   Illustration of the new basic principle of the technosystem and new technologies in it.

TABLE III

Consumers process – origin of wastes

Losses of products
Wastes produced by conversion of products
Inactivated products i.e. by dilution or contamination
Worn products
Package
Contamination of environment by use of products.

(solid and liquid) wastes as different problems. Figure 3 will point to the fact that this separating of waste problems in different kinds of problems could lead to wrong conclusions. The selection of suitable treatment processes must include the whole field of wastes produced as well as all the processes until wastes are ultimately disposed of. To convert one type of waste problem into another could be no solution at all even if it sometimes could mislead the authorities.

to drainage of these wells. The desulfurization of crudes or fuel oils could be made by a process shown in Figure 5. By high pressure, high temperature hydrogenation the sulfur compounds are transferred to hydrogen sulfide and a fuel with a reduced sulfur content is produced. At the same time there is a certain loss of heating value of the fuel oil. The hydrogen sulfide formed is separated from the circulated hydrogen and can be transformed into elemental sulfur by a conventional Claus process. The hydrogen consumed must be produced in special units as the quantity of by-product hydrogen of the refinery is not large enough to cover the quantity needed by a reasonably high reduction of sulfur content of the fuel oils produced in the same refinery. These processes are technically well established and full scale units are started or planned

TABLE VII

Stack gas desulfurization

Type of process	Agent	Products	Size of process
Absorption	Ammonia	Ammonium sulfate	Full scale
Absorption	Sodium sulfite–Sodium bisulfite	SO_2	Full scale
Oxidation	Air	Sulfuric acid	Pilot plant, full scale planned
Absorption	Magnesia slurry	SO_2/Sulfuric acid	Full scale planned
Absorption/Electrolysis	Caustic	SO_2 + hydrogen + oxygen	Commercial test planned
Adsorption	Limestone powder	Calcium sulfate	
Absorption	Limestone slurry	Calcium sulfate	Commercial test
Reduction etc.	Potassium formate–carbon monoxide	Thiosulfate–Hydrosulfide – H_2S	Pilot plant planned
Absorption	Copper oxide	SO_2	Commercial scale planned

all over the world. The investment costs are high and the desulfurization costs per unit sulfur are highly depending upon the income of the sulfure produced i.e. sulfur market price and transport costs.

Another way of tackling the sulfur problem is to reduce the content of sulfur dioxide in the flue gases of power stations etc. As can be seen from Table VII several types of processes have been suggested and tested in small scale but very few are in full scale operation. Technical and economical problems have prevented their wider use. The stack gas purification either leads to products of limited value, eventually with new problems of waste disposal, or requires additional units for further treatment, e.g., into elemental sulfur. Only in some special cases are, however, these units of a size big enough from technical and economical point of view.

7. Coal

In Sweden and some other countries coal plays an insignificant role among the energy resources; petroleum fuels dominate. In U.S. and many other countries coal is the dominant fuel. 68% of the sulfur dioxide emitted in U.S. last year is produced from coal. The coal reserves with low sulfur content are situated far from the centers of population and industry, which means high transport costs. Several processes to reduce sulfur content of coal-based fuels have been investigated but so far without significant success. Such processes as extraction of pyritic sulfur from coal and conversion to liquid or gaseous fuels combined with sulfur removal have been tested. Flue gas purification in power stations etc. is of course also applicable to coal fuels and offers the same problems as by petroleum fuels. In U.S. reports there is very little hope for a significant reduction of the sulfur from coal in the next decade. It is very important, however, that this problem is solved if coal is to take over the future energy production when oil reserves are reduced.

8. Sulfur Market

The technical problems of fuel oil desulfurization are solved to-day and the by-product sulfur is a well known and widely used raw material. The economy of the process is, however, as mentioned before, highly depending on the market price of sulfur. The amount of fuel oil sulfur will be a very big one compared to other quantities in this market if a high reduction of sulfur content is accomplished.

In Sweden the total sulfur content of consumed fuel oils is about 85% of all sulfur consumed. More than 50% of this consumption is connected with pyrites and cannot be replaced by fuel oil sulfur without affecting another raw material market. If only elemental sulfur and sulfur dioxide are replaced by fuel oil sulfur this amount will

TABLE VIII

Approximate sulfur production costs
($ per long ton sulfur equivalent)

		Manufacturing cost	Return on investment	Sales, general, and administrative	Minimum FOB price
Frasch	Low	7	3	4	14
	Medium	11	4	4	19
	High	15	8	4	27
Sour gas	Natural gas	8	7	2	17
	Refinery	15	5	2	22
Smelter gas		6	12	2	20
Pyrites		20	15	3	38
Gypsum		25	10	3	38
Other native		25	10	3	38
Utility stack gas		25	18	3	46

TABLE IX

Free World sulfur demand (million long ton per year)

Consumption	1970	%	1975	%	1980	%
Sulfuric acid						
Phosphate fertilizer	10.0	36	13.5	37	17.2	38
Ammonium sulfate	1.5	5	1.65	5	1.8	4
Other (refining, TiO₂)	12.4	44	15.8	43	20.0	43
Sub-total	23.9	85	30.95	85	39.0	85
Non-acid (CS₂, pulp and paper)	4.3	15	5.5	15	7.0	15
Total	28.2	100	36.45	100	46.0	100

TABLE X

Free World sulfur supply (million long ton per year)

Production[a]	1970	%	1975	%	1980	%
Frasch	8.3	28	11.5	26	15.5	25
Sour natural gas	7.2	24	12.0	27	18.5	30
Smelter gas	2.8	9	4.0	9	5.7	9
Petroleum	1.9	6	3.6	8	6.1	10
Utility stack gas	0.1	1	0.6	1	1.6	2
Pyrites	7.0	24	9.0	21	11.0	18
Other	2.2	7	2.8	6	3.6	6
Total	29.5	100	43.5	100	62.0	100

[a] If current trends were to continue.

correspond to a reduction of fuel oil sulfur content of about 55% or a reduction from 2.5% to about 1% in the heavy fuel oils. In U.S. 0.3% represents the limit demanded by certain states, e.g., by New York from October 1, 1971.

When we look at a bigger market as shown in Tables VIII–X the problem is much more complicated. Table VIII shows the FOB prices of sulfur from different sources. It is essential to notice the facts that Frasch sulfur is one of the cheapest sources of sulfur and that refinery sulfur probably has to suffer from high transport costs. The sulfur recovery projects for petroleum fuels planned two to three years ago were based on a price of 40 dollar/ton sulfur. The transport costs can play an important role in determining these market prices.

As shown in Table IX the Free world demand of sulfur is expected to grow from about 28 million ton/year in 1970 to about 46 million ton/year in 1980. According to Table X the supply is expected to grow during the same period from about 30 million ton/year to about 62 million ton/year. 1980 there will be an overproduction of about 16 million ton/year. In the forecast considerably less than 50% of petroleum fuel sulfur and less than 10% of the stack gas sulfur is recovered. In this market a FOB price of about 10 dollar/ton refinery sulfur is more realistic than the 40 dollar/ton actual a few years ago.

Obviously a reduction of sulfur emission acceptable from environmental point of view cannot be reached without seriously affecting the sulfur market. The Frasch sulfur is the only supply which could be reduced without affecting the supply of other elements or the pollution with sulfur compounds. Frasch sulfur is, however, one of the cheapest sulfur supplies and not big enough to correspond to the sulfur produced if all fuel sulfur is recovered. Should the Frasch sulfur supply be excluded by legislative means and saved for future use, should elemental sulfur from petroleum fuels be stored as such and/or stack gas sulfur be deposited as harmless compounds such as calcium sulfate? Solving of such types of problems is essential to the future reuse of wastes. The right solutions can only be achieved by cooperation between different cathegories of experts, by cooperation between industry and community and by taking into consideration the whole system of the human activities and the natural resources included in or affected by these activities.

The extensive work on principal problems of waste reusal shall not, however, exclude special or local solutions of waste problems. An excellent example of such a solution was recently published by the Boliden Co. in Sweden. The calcium sulfate obtained as a by-product in their new phosphoric acid plant will be sold to the cement industry and other industries producing building materials. The production of calcium sulfate will be about 500000 ton/year compared to a consumption of about 700000 ton/year. Their new sulfuric acid plant and phosphoric acid plant will produce a sulfate pure enough to fulfil the requirements of the demanding industry.

9. Conclusions

I would like to finish this rhapsody on waste problems by stressing:

(a) the need of principal solutions as well as special ones considering the effects on all parts of the total system,

(b) the need of cooperation between different kinds of scientists and experts as well as between industry and community,

(c) that engineering societies hold a special responsibility for contributions to this cooperation and to the successful results achieved in such a cooperation.

The choice of subject of this publication should be considered as one of the signs of the fact that the scientists, experts, and societies are ready to take an active part in this work. Another sign is the committee of waste treatment formed in cooperation between the Swedish academy of engineering sciences and the Swedish engineering society.

At last a picture may be shown supposed to illustrate modern life, showing an autocar wreck in frcnt of a modern appartement building and with the following caption:

This is a monument of welfare. The donator has a definite wish to be unknown.

One of my desires for the future is that there will be no need for publishing a picture like that, e.g., with a pile of drums containing dangerous chemical wastes in the foreground and an empty building in the background. We have technical ability and economic resources enough to avoid that. Let us only use them in a right way.

THE DEVELOPMENT IN THE CHEMICAL INDUSTRY
WITH AN EMPHASIS ON AIR AND WATER POLLUTION

PER SÖLTOFT

Technical University of Denmark, Lyngby, Denmark

Abstract. The growing of the chemical industry is briefly outlined with its problems of air and water pollution. Different methods for the abatement of pollution are discussed to reach the goal of constant pollution as a minimum. All methods are illustrated by examples.

For many years it has been necessary in the chemical industry to pay attention to the health hazards which are involved in the utilization of chemical compounds. A great number of intermediate compounds are ill-smelling, poisonous, or caustic and many solvents give off obnoxious vapours. Harmful effects among the employees are nevertheless rather scarce which fact is probably due to the careful supervision carried out by the authorities as well as by the industrial concerns.

In recent years an ever increasing number of warnings are heard maintaining that the chemical works have a harmful effect on their environment. The waste products leaving the factories through the atmosphere, through the waste water, or as more concentrated solid waste, exert an influence on the environment, and although major acute cases are relatively infrequent it must be assumed that by a prolonged exposure at a lower concentration level they contribute to a depreciation of the conditions of life for man, animals and plants. The relevant hazards must be looked upon by the community with considerable anxiety and in all circles people agree that it is necessary to make a serious effort to reduce the pollution.

The fact that the problems have grown so large must be seen as a result of the fast development of industry. The demand of modern man cannot any longer be satisfied with food, clothes and dwellings only. To an ever increasing degree it includes also a long series of industrial products. Countries like Denmark which were formerly dominated by agriculture are more or less forced to develop industry, and among industries the chemical industry is a branch undergoing a particularly fast development.

On a global basis the chemical industry has developed from a modest size in the twenties, when the production value was about 5×10^9 \$ to about ten times as much around 1950. To-day it has again been tripled so that the production value amounts to about 170×10^9 \$ (Figure 1).

The pollution caused by this remarkable expansion has attracted much attention and unfortunately it has the effect that industry is becoming rather unpopular. A large part of the population has attained a hostile attitude towards industry. Some may even be of the opinion that it would be better to dispense with it. Against such an opinion it must be stated that by far the largest part has been developed in order to cover the demands of the consumers. It has in fact been a wish to have it, and the

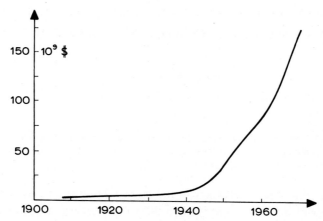

Fig. 1.   Chemical world production (from Ullman: *Encyclopädie der technischen Chemie*, 1970).

industrial products contribute to a large extent to the pleasure and comfort of our daily life. It is too one-sided to consider only the inconveniences, and a more balanced view should be taken.

But still the pollution must be combatted. The minimum requirement which must be fulfilled is that the pollution should not be allowed to increase even though the amount of industry is increasing. Of course it is still better to improve conditions but this will be feasible only if more severe standards are accepted for the amounts of harmful compounds allowed in the discharges from the chemical works. This again will make it necessary for the chemical works to develop improved methods to avoid pollution. In this connection it might be useful to consider a number of possibilities.

(1) Alternative methods of production.

(2) Improvement of the degree of conversion.

(3) Disposal of by-products by recirculation.

(4) Disposal of waste water.

(5) Disposal of waste gases.

1. Alternative Methods of Production

Economic considerations have always been very important for the selection of a method of production if the desired product could be made in several ways. This is still true, but as an additional important factor, possible differences in pollution effects are now taken into account. As an illustration of this kind of problems a few examples from recent experiences will be mentioned.

Phenol, a chemical used in vast amounts, may be produced by several processes (Figure 2).

By the classical sulphonation of benzene the excess sulfuric acid appears as an unpleasant by-product and from the alkali fusion also sodium sulphite is obtained. By the chlorination of benzene hydrochloric acid is formed in the first process and

PHENOL

Sulfonation of benzene

$$C_6H_6 + H_2SO_4 \rightarrow C_6H_5SO_3H + H_2O$$
$$C_6H_5SO_3Na + 2NaOH \rightarrow C_6H_5ONa + Na_2SO_3 + H_2O$$

Chlorination of benzene

$$C_6H_6 + Cl_2 \rightarrow C_6H_5Cl + HCl$$
$$C_6H_5Cl + 2NaOH \rightarrow C_6H_5ONa + NaCl + H_2O$$

Oxidation of cumene

$$C_6H_5CH(CH_3)_2 + O_2 \rightarrow C_6H_5(CH_3)_2OOH$$
$$C_6H_5(CH_3)_2OOH \rightarrow C_6H_5OH + (CH_3)_2CO$$

Fig. 2. Phenol production, different methods (from Ullman, *Encyclopädie der technischen Chemie*, Ergänzungsband 1970, p. 183).

common salt in the following. One of the more recent methods starts from cumene. This is oxidized without any by-products to form cumene peroxide which is finally split into phenol and acetone, two products both of which can easily find a market. From a pollution point of view the cumene method is therefore very favourable, and in Western Europe about 88% of the 1968 production was produced by this method, which also seems to be the preferred method for practically all new plants.

Propylene oxide is normally produced by a process in which propylene is first chlorinated to propylene chlorohydrin (Figure 3) followed by a treatment with lime slurry to give propylene oxide.

As by-products hydrochloric acid as well as calcium chloride are produced. It would be natural to try to evade this by use of a direct oxidation of propylene with air but this process has not yet been fully developed for industrial use. The Halcon

PROPYLENE

Classical method

$$CH_3CH = CH_2 + Cl_2 + H_2O \rightarrow CH_3CHOHCH_2Cl + Cl^- + H^+$$

$$CH_3CHOHCH_2Cl + \tfrac{1}{2}Ca(OH)_2 \rightarrow CH_3\overset{O}{\overset{\diagup\diagdown}{CH.CH_2}} + \tfrac{1}{2}CaCl_2 + H_2O$$

Halcon method

$$(CH_3)_3CH \xrightarrow{O_2} (CH_3)_3COOH + C_3H_6 \rightarrow (CH_3)_3COH + CH_3\overset{O}{\overset{\diagup\diagdown}{CH - CH_2}}$$

$$(CH_3)_3COH \xrightarrow{\div H_2O} (CH_3)_2C = CH_2 \xrightarrow{H_2} (CH_3)_3CH$$

Fig. 3. Production of propylene oxide (from Kirk-Othmer: *Encyclopedia of Chemical Technology* **16**, 1968, p. 603).

method follows an indirect course. Iso-butane is first oxidized to a peroxy compound and this is then reacted with propylene to form propylene oxide and trimethyl carbinol. From this alcohol iso-butane can be regenerated by dehydration and hydrogenation. It is seen that this process can be carried out without any formation of by-products.

2. Improvement of the Degree of Conversion

It is common practice to endeavour to utilize the raw materials to a large extent in all chemical reactions in the production line. However, it is difficult always to get 100% conversion. The non-reacted raw materials may under certain circumstances pass through the plant and contribute to the pollution. In such cases the situation can be ameliorated by an increase in the degree of conversion. At the same time the production will give a larger yield. However, it is not certain that it will give an improved economic result as the increase in conversion may be connected with a considerable increase in cost.

A typical example of this kind of problem is the catalytic sulphuric acid production. In this process sulphur dioxide is oxidized to sulphur trioxide but in a conventional plant it is normally not possible to convert more than 98%. Therefore a certain amount of sulphur dioxide is always found in the tail gases leaving the sulphuric acid plant. The concentration may be estimated at about 1000 ppm.

In Bayer's Double Contact Process (Figure 4) an extra absorption of sulphur trioxide is applied already when about 90% of the sulphur dioxide has been oxidized.

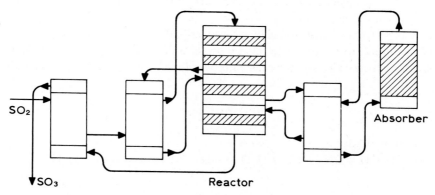

Fig. 4.   Bayer process, simplified flow sheet (from Ullman: *Encyklopädie der technischen Chemie*, **15**, 1964, p. 459).

After this absorption the gas stream with the remaining sulphur dioxide is returned to the reactor where it passes an additional layer of catalyst. The concentration of sulphur trioxide is now very low, and the reaction of the sulphur dioxide is thus facilitated so that the total conversion can reach about 99.8%. It is seen that by this process the amount of unreacted sulphur dioxide has been reduced to one tenth. The yield of sulphuric acid has been increased by about 2% but to get this increase it has been necessary to make use of a more complicated and more expensive plant.

An increase in the degree of conversion may also be obtained in a conventional plant by use of an improved catalyst which allows the reaction to take place at a lower temperature. By some practical plant experiments very encouraging results have been obtained with a recently developed Danish catalyst.

3. Disposal of Byproducts by Recirculation

Unfortunately in chemical production it is not always possible to prevent the formation of by-products which must be disposed of. In such cases it should primarily be investigated whether the by-products could be isolated and eventually recirculated into the process. This is a classical type of problem which already for economical reasons attracts great attention and the history of chemical engineering is full of examples.

By the earlier mentioned method for the production of phenol by chlorination of benzene the byproduct hydrochloirc acid is formed. In a modification of this process, the so-called Raschig-Hooker process (Figure 5) the chlorination is carried out with a mixture of hydrochloric acid and air and in this way the hydrochloric acid is reused in the process. At the same time the alkaline hydrolysis is replaced by a hydrolysis with water vapour at high temperature so that theoretically no waste products are formed.

PHENOL

Raschig-Hooker Process

$$C_6H_6 + HCl + \tfrac{1}{2}O_2 \rightarrow C_6H_5Cl + H_2O$$
$$C_6H_5Cl + H_2O \rightarrow C_6H_5OH + HCl$$

Fig. 5.   Phenol production, Raschig-Hooker process.

However, it is not by far possible always to recirculate the by-products. As an example from the inorganic industry the production of soda ash by the ammonia soda process may be mentioned. Important amounts of calcium chloride solution are obtained and must be cleared away. Also in organic production the destruction of the by-products often cause serious troubles because many industrial concerns do not have the necessary installations available. In several countries a preparatory work is just now going on to organize the collection and incineration of industrial waste. This is expected to be of importance especially for industrial concerns of a moderate size.

4. Disposal of Waste Water

An important type of industrial pollution is connected with the waste water. In many instances the amount of waste water can be kept down by selection of appropriate apparatus.

In scrubbers the waste water can be recirculated and it may be possible in this way to obtain an increased concentration of the extracted compound sufficiently high for an isolation to be economically feasible. In cases where steam distillation was formerly used and a contaminated water phase was therefore obtained, vacuum-distillation has now been substituted and water-free products are obtained. Where impurities were formerly removed by washing with watery solutions, distillation is now often used for the purification and the impurities are isolated in a concentrated state ready for incineration. Barometric condensors in connection with vacuum plants yield huge amounts of polluted water and can with advantage be changed to surface condensors in which the water quality is not damaged.

Strict regulations must to-day be observed by the concern that wishes with confidence to be able to make use of a lake or other sink as a recipient for waste water. It may therefore happen that it will be found economical to treat the water with the object of reuse in the process rather than to let it out into the sink.

The amount of cooling water from a plant can be reduced by use of closed systems in which the hot spent water is cooled in cooling towers for reuse. Waste water is eliminated and the consumption of fresh water is reduced to the amount necessary to make up for evaporated water. In many modern plants water cooling is avoided and advantage is taken of air cooling in large finned tube coolers.

A marked difference exists of course between water which has been in contact with chemicals and water which has only been used for cooling purposes. In many cases it will pay to provide for a double sewage system so that the amount of water needing purification is substantially reduced.

5. Disposal of Waste Gases

Atmospheric pollution from chemical industries is another field in which a substantial amount of work is being done. Dust is fought by more and more widespread use of well-known apparatus like cyclones, ordinary filters, electrostatic precipitators and scrubbers. The tail gases leaving a chemical plant will often contain contaminating chemical compounds which must be removed. Hydrochloric acid is often formed by chlorination of organic compounds and must be removed by absorption, but in this case unfortunately the purification of the air gives rise to the occurance of a water pollution problem. Such interdependence of pollution problems is quite common. Hydrogen fluoride from the superphosphate production can fortunately be utilized in the production of sodium fluosilicate, a commercial product which may contribute to making the purification process profitable.

A couple of air pollution problems have attracted special attention, viz. sulphur dioxide in the tail gases from the production of sulphuric acid and nitrogen oxides in the tail gases from the production of nitric acid. The development work being done to diminish the outlet of sulphur dioxide by improvement of the conversion in the sulphuric acid plant has already been described. The emission of nitrogen oxides looks very dramatic because the oxides appear with a strong reddish brown colour. It can be observed in many places as the total production of nitric acid amounts to about 5 million ton per year. In normal production the nitrogen oxides are absorbed in water but although this absorption is presently carried out at increased pressure it is not possible to obtain colourless gases by this means alone. Other absorption liquids give a more efficient absorption but tend to create problems in connection with the disposal of the spent liquor.

In one of the recent methods the nitrogen oxides are reduced with natural gas (Figure 6). A stream of gas is added and a part of the hydrocarbons react with the

$$CH_4 + 4O_2 \rightarrow CO_2 + 2H_2O$$
$$4NO_2 + CH_4 \rightarrow 4NO + CO_2 + 2H_2O$$
$$4NO + CH_4 \rightarrow 2N_2 + CO_2 + 2H_2O$$

Fig. 6.   Reduction of nitrogen oxides.

content of oxygen while another part reduces the nitrogen oxides to free nitrogen. A platinum catalyst is used. A satisfactory reduction is obtained only when an excess of gas is used. A similar method has been proposed in which the reduction is carried out by means of ammonia using again a platinum catalyst. None of these methods have so far been satisfactorily developed and they have not found any extensive use in practice. They have both been described as economically unfavourable. However, a further development in this field can probably be expected in the near future.

It is not to be expected that chemical works could be freed from all nuisances. It is therefore important to be careful in the selection of a proper site. In this connection many conditions have always been of importance. The site should be favourable for transportation of raw materials and products. Water, electricity and labour should be easily available etc. In the future it is necessary to add to this the important considerations concerning disposal of effluents and possible harmful effects on neighbouring people caused by contaminated air, by odour or by noise.

The various pollution problems will require a great effort for their solution. Work must be done on the alteration of production routes and the improvement of apparatus. Considerations are needed concerning the treatment of waste products and the favourable location of chemical works. All these tasks are typical chemical engineering problems. Therefore a heavy responsibility rests on the chemical engineers, a responsibility of which most chemical engineers are well aware and for which they are willing to work.

The many pollution problems must be taken into consideration when advantages and disadvantages by the operation of a new chemical plant are compared. Most obvious is it that the plant should allow a satisfactory economic result. At the same time it will also contribute to employment and normally it will make available a number of desirable products. For a country like Denmark where the number of people living from agriculture has been halved within the last 20 years the growth of industry has been an indispensable factor in employment. Against these beneficial effects must be weighed the possible harmful effects which the plant could exert upon the health conditions of the population, upon its comfort and more generaly upon the environment.

The balancing of advantages and drawbacks is complicated by the fact that no common measure exists. How could the possible damage to the public health be stated as a sum of money? It is in reality a political problem to decide how far a chemical concern should go with regard to investment of money for improvement of relations to the environment. In recognition of this fact the government of most countries have set up regulations for the maximum allowable pollution level and in countries where this has not yet been established action will most certainly soon be taken. Everybody realizes that such regulations can be considered a safeguard not only for the population but also for the industry.

In the discussions of the influences of the chemical industry in recent years a tendency can be seen to lay considerable stress on the pollution effect. Industry has been described as entirely or at least predominantly undesirable which is of course not correct. Some have imagined that the overwhelming part of the pollution originates from industry. It cannot be denied that some types of pollution are characteristic for the chemical industry.

In other cases the chemical industry occupies only a position among other sources of pollution.

In an American specification of emissions from 1968 industry has been compared to three other sources of pollution; stationary combustion sources, mobile combustion sources, including inter alia automobiles and incineration plants for solid waste (Table I). It is seen that the contribution of industries for particles amounts to

TABLE I

Estimated emissions in United States (from Hoffman A., National Air Pollution Control Administration: National Emission Reference File, 1970).

Estimated emission in U.S.A. 1968, mill. ton	Particles	SO_2 SO_3	NO_x	CO	Hydrocarbons
Stationary combustion	8.1	22.1	9.1	1.7	0.6
Mobile combustion	1.1	0.7	7.3	57.9	15.1
Incineration	0.9	0.1	0.5	7.1	1.5
Industry	6.8	6.6	0.2	8.8	4.2
Total	16.9	29.5	17.1	75.5	21.4

about 40%, for sulphur dioxide to about 22%, and for nitrogen oxide to about 1%. It should be added that in Denmark where the chemical industry is relatively less developed than in the U.S.A. the contribution presumably amounts to much less. Sulphur dioxide is probably only some few per cent. As a curiosity may be mentioned that the tail gases from a sulphuric acid plant of the tower system, a common system in Denmark, will contain normally only about 0.2 to 0.3 gram of sulphur compounds per cubic meter while stack gases from a power plant in which fuel oil with 2.7% sulphur content is used will contain about 4 gram of sulphur dioxide per cubic meter. On the other hand it is true that industry must be very careful to take precautions against irregularities in the production which might give rise to considerably larger concentrations.

In summing up it should be stated that the purpose of this paper is to show that a certain part of the pollution of our environment originates from the chemical industry. It is further mentioned that it is possible to a large extent to counteract this pollution already in the plant where it is formed. A brief description of the principles for this fight against pollution is given and a number of examples are mentioned in which good progress has already been made or promising ideas have been presented.

It is underlined that research work in this field is of importance because the chemical industry must be considered an indispensable part of modern community. Fortunately there is every reason to believe that we shall soon see useful results of the many efforts which are now being made.

LOCATION, SIZE, AND INTERACTION OF CHEMICAL PLANTS

H. W. KNAUFF

Director, Bayer AG, Leverkusen, F.R.G.

Abstract. Different factors governing the location of chemical plants are discussed. The influence of other factors on determining the appropriate size of a chemical plant is described. Interaction of such plants, especially cooperation within one industrial complex, offers important economic advantages.

Before tackling the problems of location, size and interaction of chemical plants, I want to state that in this context the term 'chemical plant' means a plant for the synthesis of a product by one or more chemical reactions but not a processing plant e.g. for the finishing of pharmaceuticals or the formulation of insecticides. While for the latter type of plant location and size are certainly no negligible factors, the three parameters in the heading of my paper are, for many real chemical productions, the decisive factors determining the fate of an existing plant in a highly competitive situation or of a project to build a new plant.

1. Location of Chemical Plants

The choice of the proper location depends upon several factors, mainly utilities, availability of raw materials, cost and time of transport of finished goods to the customers, disposal of effluents, availability of labour and key-personnel.

Let us first consider the influence of utilities.

Every chemical plant requires power, water, and fuel, many plants require also gas. When planning a new large chemical grass roots plant one can assume, based on average consumption figures in large chemical plants, that per hectar of built up land an installed capacity, in the final phase, of 10–15 t/h steam, 500–1000 kW and 100–200 m^3/h water will be required.

1.1. POWER

In the highly industrialized countries power is nearly everywhere available from the grid; in developing countries availability of power may be a problem. For many productions cheap power is imperative as will be seen from Figure 1.

Large electro-chemical or electro-thermal plants are hopelessly uncompetitive at locations with unfavourable power supply. Such plants are mostly located in areas with cheap hydroelectric power or with power generated from cheap lignite or hard coal e.g. the Alps, Norway, Central Germany or the Rhine/Ruhr area. Due to the inroads made by fuel oil as an energy carrier the picture is changing; Hoechst's new phosphorus plant has been built at Flushing. With the growing use of fuel oil and when atomic power will become more competitive the choice of location even for chemical plants with important power consumption will become easier.

G. Lindner and K. Nyberg (eds.), Environmental Engineering, 43–52. All Rights Reserved

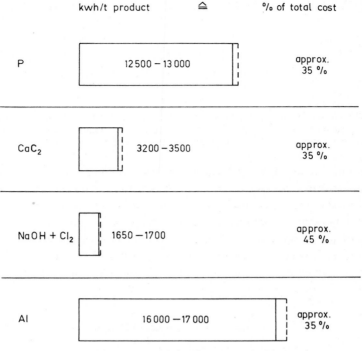

Fig. 1.   Power cost in % for different chemical industries.

In principle, any well-sized chemical plant may also generate itself the power it needs. In general, this is only economical if power obtained by back pressure is a by-product of steam generation. A chemical plant will therefore only make as much steam as it requires itself, obtain the equivalent of power by pressure release to the pressure needed for its manufacturing processes, and buy the balance of required power from the grid. Generation of the relatively small amounts of additional power needed by condensation of steam would be uneconomical compared with buying it from the large-scale specialized power plants.

1.2. STEAM

Steam cannot be transported over great distances; it has to be generated near the spot of consumption. Its price largely depends upon the price of fuel and its handling cost. Figure 2 shows how the prices of fuel oil, hard coal and lignite, expressed in DM/Gcal, developed after 1945. It explains why areas with abundant hard coal deposits such as the Ruhr area have lost much of their attractiveness as sites for chemical plants.

Fuel oil has to a considerable extent replaced coal. Unfortunately its price has increasingly become subject to political influences. Governmental authorities in the producing, the processing, and the consuming countries want to levy as much taxes thereon as possible and sometimes even restrict exploitation. Freight is an important

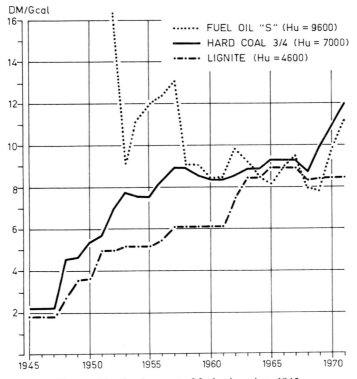

Fig. 2.   The development of fuel prices since 1945.

cost factor so that in a free market fuel oil is the cheapest in the neighbourhood of a refinery, particularly a refinery close to a seaport for large tankers.

1.3. WATER

Fresh water is a precious good in many countries. Though nowadays the requirements of cooling water of a chemical plant may be considerably reduced by air cooling and recycling, or sea-water may be used for this purpose under special precautions, fresh water is still an indispensable utility as process water. Non-availability of sufficient quantitites of fresh water is therefore a limiting factor in the choice of location. It narrows the choice e.g. in hot countries such as India, Pakistan, or part of Spain and Southern Italy. Even at otherwise well suited places on the North Sea coast as in Antwerp fresh water is not in plenty supply and therefore costly.

1.4. TRANSPORT

Once a number of sites with suitable utility supply have been found the problems of raw materials supply and transport of the finished goods have to be studied.

If large quantities of minerals are needed, freight plays an important role. This is one of the reasons why e.g. a number of titanium dioxide plants were erected near seaports. Freight cost for large quantities of basic cheap chemicals such as sulphuric or hydrochloric acid, chlorine or hydrogen may either exclude a special site or render their production on the spot necessary, which increases the investment cost and risk. For all its TDI plants outside Germany, Bayer* had to look for sites with adequate cheap hydrogen supply from outside for the hydrogenation of dinitrotoluene since generation of hydrogen would have rendered the project uneconomical. Petrochemical activities should be close to a cracking unit unless a pipeline system for the basic olefins is available. The location of units for the manufacture of relatively high priced goods such as pharmaceuticals is, of course, less dependent on a nearby source of supply of basic raw materials.

With regard to the disposal of the finished goods the choice of the site depends on whether the plant will be destined to satisfy mainly the demands of the economic area in which it is built or whether it will be preponderantly orientated towards sales over great distances such as exports to overseas countries. In the first case the ideal location would be in the center of consumption. If the production is mostly sold in far-away countries a manufacturing site near an important seaport should be chosen unless it appears better to manufacture in one of the overseas countries itself which consumes a good deal of the envisaged production.

1.5. EFFLUENTS

Disposal of effluents now plays an important role in the choice of a suitable site for the synthesis of chemical compounds on an industrial scale. Suitable effluent treatment should nowadays form part of every manufacturing process. In addition thereto, all large chemical plants should have central effluent treatment plants. Since the treated waste water has to be disposed of, every large chemical plant will be located near a big river or at the seaboard so that their waste water is quickly and sufficiently diluted. Dumping of residues into the open sea, where permitted, is, of course, cheaper from a coastal than from an inland site.

The composition of the effluents plays an important role. While many organic by-products can be biologically decomposed with the formation of innoxious compounds, the elimination of many water-soluble inorganic compounds at reasonable cost causes moie problems. Plants generating large quantities of e.g. $NaCl$ or $CaCl_2$ should be located as closely as possible to the sea and not the source of a river.

Air pollution, which to a certain degree is inevitable, even though economically bearable effective measures to reduce it to a minimum are taken, excludes the construction of new big chemical complexes within densely inhabited areas, especially those leading to smog formation. The prevailing wind direction has to be taken into account. From the standpoint of air pollution a site at the seaboard with fresh coastal winds is advantageous.

It should also be considered as to whether possible own air pollution may disturb the pioduction of others or whether possible pollution from outside may cause trouble

---

* Bayer AG, Leverkusen.

to the envisaged own production. It would, for instance, not be wise to build a new photographic film factory near a crude oil refinery emitting, in case of production troubles, volatile sulphur compounds.

1.6. LABOUR

Availability of good qualified labour has become a problem in many parts of the world. In countries with a considerable labour shortage such as Western Germany industry increasingly tends to export jobs to countries where labour is relatively abundant and cheap instead of importing more foreign labour. Labour shortage was one of the reasons why Bayer built and is building chemical factories e.g. in Belgium and Spain and installs processing plants all over the world so that right now Bayer has about 85 operating plants outside Germany. The tremendous wage increases asked for and granted to the powerful trade unions in many industrialized areas of Europe during the past years will certainly increase this trend towards multinational production.

1.7. KEY PERSONNEL

No plant will make profit without good key personnel. Conditions of life, including educational and cultural facilities at or near the location of the plant, should be such as to satisfy good key personnel. High salary alone is not always sufficient.

1.8. GOVERNMENT INFLUENCE

For obvious reasons governments and local authorities try to influence the investor's choice of the site by promising investment subsidies and fiscal advantages. One should however, not forget that they are only given to compensate local disadvantages and for a limited time only while natural disadvantages, such as lack of water, insufficient infrastructure, or large distance from the market, may last much longer.

2. The Influence of Size

As to the size of a plant it should be, generally speaking, such as to allow operation in a profitable way.

2.1. MARKET AND PRICE

The proper size of a new plant therefore depends on the importance of the attainable market. Possible sales in turn depend upon the cost price of the product to be sold in relation to the estimated selling price. The cost price is a function of many factors including depreciation and operating cost, in other words of the size of the plant, which closes the vicious circle.

2.2. DECISION ABOUT A NEW PLANT

Let us consider how in practice a decision may be arrived at: The research people have found a new product and application research has indicated that the product offers certain advantages over known products. A manufacturing process has been developed on a pilot plant scale. Market research comes up with a picture which shows, taking into account current prices of competing products and possible growth of the market for that type of product, the possible sales of the new product as a function of its sales price as represented, for instance, by curve I of Figure 3. The manu-

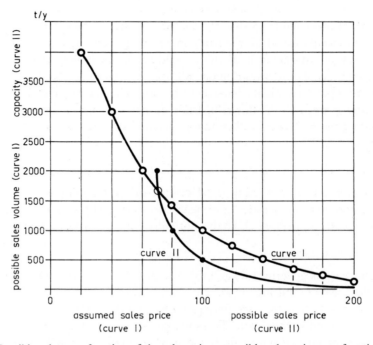

Fig. 3.   Possible sales as a function of the sales price – possible sales price as a function of size.

facturing department with the help of the engineering department will then plot a similar curve, basing itself on the result of pilot plant tests, showing the dependency of the possible sales price on the size of the operation (curve II). In this context the possible sales price is the price which would yield a reasonable return on investment which; after taxes, would be at least equal to the going bank rate.

Curves I and II cross at about 1600 t/year and a sales price of about 70. This could be considered as the appropriate size of a plant for that special product provided that the price level of competing products does not change and that curve II represents the possible sales price at around 80–90% effective use of the capacity. I would not recommend calculating with a higher use of designed capacity since allowance should be made for repair time in order to avoid too heavy wear and tear.

Unfortunately, things are not that simple. Curve I is based on the *present* price level of competing products. The marketing people may, however, predict that two years from now when the plant will be ready the price level of the competing products will be 20% down and that five years from now it may be 30% lower. Since, normally, one has to remain in business at least 7–10 years in order to amortize a plant, another pair of curves has to be drawn, taking into account a price drop of 30%. Now, the curves do not cross anymore.

Unless the possible sales price can be drastically reduced, mainly by improving the manufacturing process, the planned production would not be competitive.

2.3. EFFECT OF SIZE ON INVESTMENT AND OPERATING COST

Size has, of course, an effect both on investment and operating cost. Doubling capacity 'within battery limits' costs, as a thumb rule, in general only 1.6 times the original investment. The corresponding factor for operating cost may vary between 1.7 to 1.8.

The picture changes even more if also the cost of off-sites is considered. When building a grass roots plant one has to invest around one third of the total sum in off-sites. They are in general sufficient even for doubling the plant since it would be uneconomical to build e.g. the initial sewage system or the power plant without any provision for future expansion. The fact that effluent treatment plants with inter alia settling, treatment and retaining facilities as well as measures to avoid air pollution may considerably add to offsite costs should not be neglected.

The high investment in off-sites entails, on the other hand, the necessity to start a new grass roots factory with a relatively high production and sales volume and hence an important investment in assets within battery limits in order to get a great divisor for the general overhead. This consideration led Bayer to spend more than DM 200 million as initial total investment when starting construction of each of its two Antwerp plants on both sides of the Schelde river.

The favourable effect of size on investment and operating cost per unit produced together with very optimistic sales forecasts have led many companies to erect bigger and bigger capacities for the production of a given product. The dangers of such gigantomania should not be lost sight of: if, as a rule, the utilization of capacity of such specialized plants falls below 85–80%, operation results in a loss which, with such big plants, may be very high. In addition hereto, the tendency to pass the competitor by building a still bigger unit often leads to the creation of overcapacity and hence to a breakdown of the sales price of the product in question. The recent breakdown of prices on the fibre market is an example of the inauspicious consequences of an unconcerted capacity struggle.

With a view to ensure high utilization of capacity and to benefit simultaneously of the advantages of size, joint production subsidiaries supplying the parent companies with a given product on the basis of a 'co-operative' are increasingly gaining importance. Recent examples of such 'co-operatives' are the refinery and olefin plant at Feyzin near Lyon owned by four French chemical companies or the MDI/TDI plants which Bayer has built with Shell at Antwerp. Another possibility, mostly used by French companies, is the subscription by a company of a definite share of the capacity of a plant built by another company, the subscriber taking care of all fixed costs pertaining to his share of the capacity regardless of his off-take.

3. Interaction of Chemical Plants

A chemical plant should geographically be as close as possible to its main suppliers and customers. Since many if not most raw materials for a chemical product are chemical themselves, the source of supply of a chemical plant mostly is another chemical plant. Particularly basic and intermediate chemicals are consumed by the chemical industry itself. Among other reasons, this connection led in the past to the formation of big single, largely integrated factories such as at Ludwigshafen-Oppau, Hoechst, Billington/Wilton or Leverkusen.

Such factories produce e.g. the basic inorganic acids, chlorine, caustic soda and ammonia as well as basic organic products such as phenol, aromatic nitro and amino compounds. From such concentration there results not only a saving of freight cost: even more important is the control of the raw material supply for the finished products and the possibility of utilizing unavoidable by-products on the spot. This explains why in the past many dye-stuffs factories made their own intermediates on the site and branched off to other fields such as pharmaceuticals or pesticides in order to make use of by-products.

In the early times of polyurethane industry the cheapest possible diisocyanate was one made from o-nitrotoluene then available in relatively large quantities as by-product of p-toluene manufacture used in the dye-stuffs industry.

Another advantage in creating big single chemical factories is the saving in general overheads per unit produced, since total overheads do not proportionately increase with volume of production. On the other hand, the advantage of bigness is at least partially compensated by the difficulty of hiring the necessary labour, of its housing, its transport to and from the factory, and the danger of financial and economic repercussions in case of unrest of labour. There is therefore an optimum of concentration which, for a widely diversified chemical plant, we in Bayer consider to be approximately 10–12000 people within one administrative unit.

3.1. COMPOSITION AND COOPERATION

Competition and shortage of capital render the formation of such big factories under the roof of one single company increasingly difficult. The economic reasons in favour of concentration, however, continue to exist, This often results in a symbiosis of several plants with different ownership at a special location. Good examples thereof are the newly erected petrochemical complexes in Japan, built around a large cracking unit.

One of the many examples is the complex in the Tokuyama area. The center is the Idemitsu refinery and its petrochemical plant with its cracking plant having an initial output of 300000 t/annum ethylene. A great number of factories, not financially connected with Idemitsu, were built around the cracking unit, making a large variety of chemicals ranging from polyethylene to polystyrene, PVC, organic solvents, polyurethane intermediates and synthetic rubbers. The necessary inorganic chemicals are supplied by two members of the combine, Toyo Soda and Tokuyama Soda.

Large chemical complexes consisting of plants of different ownership, interwoven with jointly owned companies, have recently grown in Europe, e.g. in the Rotterdam and Antwerp areas. In these cases the advantages of the location were probably the main incentives for the establishment of the single plants, but the possibility of co-operation among themselves added to the decision to settle there and not elsewhere.

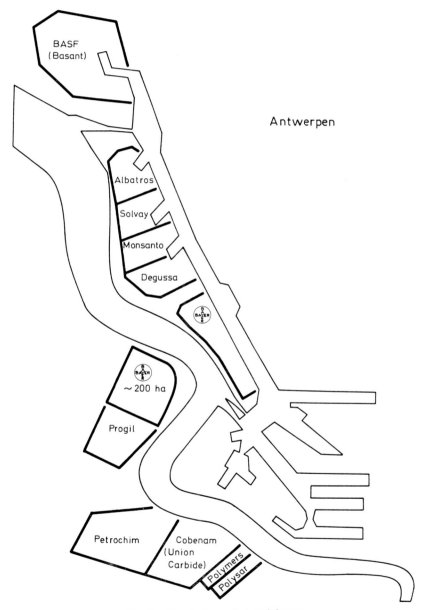

Fig. 4. The Antwerp industrial area.

Take the example of Antwerp (Figure 4). Once Bayer and BASF had bought land there, quite a number of well-known European and American chemical companies set up new plants in the area. The refineries of Esso and SIBP supply the feed stock for Petrochim's cracking plant. Petrochim's olefin production yields the raw materials for Union Carbide's polyethylene and polypropylene plants and Solvay's production of vinyl chloride and PVC. Elemental sulphur arrives from the American continent at a terminal of the Duval company which will be connected by pipeline with Bayer's sulphuric acid plant. It may also supply liquid sulphur to the future carbon disulphide plant of Progil which, in turn, will cover the CS_2 requirements of Bayer's envisaged plant for rubber chemicals on the left bank of the Schelde.

4. Conclusions

The picture drawn is one with many facets. With regard to location of chemical plants there are many advantages for coastal areas near a seaport or a site on a big river. The problem of size is very complex. It involves thorough considerations of the market situation, economic repercussions in case of insufficient use of capacity and the problem of large centralized national production versus diversified multinational production. As to interaction of chemical plants the advantages are clear; the growing shortage of capital and the need of risk sharing will in my opinion further the tendency of chemical companies co-operating in one industrial area either legally independently or on an appropriate contractual basis.

PART II

AIR POLLUTION

General

PROBLEMS OF INDUSTRIAL AIR POLLUTION

E. A. B. BIRSE

Chief Inspector H.M. Industrial Pollution Inspectorate for Scotland, Edinburgh

Abstract. The general industrial pollution problems are described. The viewpoints giving the basic information necessary for decision making of standards are outlined and industrial examples discussed. The role of the chemical engineer is given.

For industrial nations pollution problems in the chemical and allied industries have become politically sensitive in the last few years because of a sudden upsurge of public interest throughout the world in conserving the environment. While this interest was overdue, a great deal of nonsense has been put forward at times in support of conservation. Idealists have not hesitated to condemn all pollution in any form and in any amount without regard to the fact that all forms of life cause pollution, if not when living certainly when dead. Natural pollution is always with us, it takes many forms and it can be severe. It has been estimated, for example, that the volcanic eruptions at Krakatoa at the end of the last century emitted at least as much particulate matter into the atmosphere as all man made emissions to date. The capacity of the biosphere to absorb the insults of pollution is, however, not unlimited. With expanding economic growth and greater technical sophistication as universal targets, man must take heed now to ensure that his pollution does not reach such a scale and character as to interfere with natural cyclic processes in the biosphere to his own grave detriment either by seriously upsetting the normal pattern or even worse terminating them. In this respect air, water and land pollution are often indivisible since each can interact on the other.

Modern man could not live without the products of the chemical and allied industries. That is a fact of life. Another fact is that none of these industries can be introduced into a new environment without impairing the quality of its air, water and land. The task facing chemical engineers is therefore to find and apply the best practicable means to keep the damage to a minimum. There is no finality to this task. As technology advances so do the means to reduce the effect of waste discharges on the environment. These means are not confined to better ways of treating the discharges. Traditional ways of manufacture may have to be replaced by others which create less waste or waste in a form which is more readily treated. No new product should be introduced until effective ways of treating waste from its process of manufacture have been devised, and likewise the product itself must not harm the cyclic processes of the biosphere when discarded as waste by the user. Again when a chemical engineer is choosing between one process and another, the cost of treating waste discharges has become more and more important as an economic factor in the overall assessment. Cost of treating discharges may also determine the maximum economic size of any plant unit.

Ideally waste discharges should have an imperceptible effect on the environment.

What, however, is imperceptible? Just as technology advances so do scientific means of detection become more and more sensitive. Unfortunately, there is currently a tendency in some quarters to regard whatever is measurable in the environment as harmful pollution. This attitude has led to the occasional absurdity of specifying standards of environmental quality which are better than Nature provides for us. Chemical engineers need quantitative guidance from the biological sciences to assist them in determining the degree of treatment required for waste discharges but most of these sciences have some way to go before they can do so with sufficient precision for design purposes. In the meantime, chemical engineers have an inescapable responsibility to take the best practicable means currently available to them to reduce to a minimum damage to the environment by waste discharges.

1. Control of Industrial Emissions

The chemical and allied industries have few, if any, air pollution problems which are incapable of solution to the point where the discharge pollutants are acceptable and harmless to the public in general and their food supplies. Ample technical knowledge is available to achieve high efficiency in absorbing gases, in arrestment of grit and dust, in prevention of smoke emission and in discharging the inevitable residues so that they disperse well in the atmosphere. The basic problem is one of finding capital for the necessary waste treatment plants and the revenue to run the plants. The solution to this problem turns on being able to persuade product users to pay the cost of waste treatment in the price of the products. National legislation on pollution control has a part to play in this matter to ensure a reasonable degree of uniformity within any industry. Tempting though it may be, however, it is an over-simplification to extend this argument to seek complete uniformity of dischargd standards on the ground that the cost of pollution treatment should be equally shared within an industry. An existing plant unit which is old but still has an economic life may not be capable of being modified to meet in full a new universal discharge standard. On the other hand a new plant going into an environment of high amenity value may require to observe a more stringent standard.

After all, many other costs in an industry are not equal between the units within it, such as transport, water costs, fuel and so on. From long experience of pollution control I am convinced that a flexible method of setting legislative standards has many advantages, and in practice it is acceptable to industry so long as it is seen to be applied fairly. As technology advances, old plants are replaced by new ones and at the same time discharge standards can be improved in the light of experience. Legislation which controls pollution at its source is, in my experience, effective. Plant designers and operators know exactly what is expected from them and their performance in observing source standards can be readily monitored. Control of pollution by ambient air quality standards does not seem to me to be so readily enforced as discharge standards, and if they are applied to control all discharges in an area, the burden of observing them could vary unfairly from one industry to another. At the same time ambient air quality standards are very useful as a guide in deciding standards for source control. But in all too many cases the necessary medical, biological or meteorological data for arriving at meaningful air quality standards are not available, and as a result such standards have too often to be set on an arbitrary basis. Basically, of course, reduction of pollution will always depend on action to control waste discharges at their source.

In deciding the best practicable means of reducing a waste discharge the first criterion is to ensure that there is no danger to the health of residents in the area and only thereafter is the cost of providing and operating the necessary plant taken into account. So far as health is concerned, one way of deciding on a limiting concentration at ground level is to take the threshold limiting concentration for safety of industrial workers on an 8 hour day basis, divide this figure by a factor of 3 to allow for 24 hour exposure of residents and divide again by an arbitrary factor, normally not more than 10, to allow for the frail, the very old and the very young in the resident population. Having thus arrived at a safe ground level concentration, one of the recognised diffusion formulae can be used to calculate the permissible limit of discharge at the stack. Meteorological conditions affecting diffusion from the stack vary greatly but unless particular local circumstances call for a more close analysis it usually suffices to use averages for the meteorological parameters.

Health overrides considerations of cost but in cases where discharges present no particular danger to health a decision on what is the best practicable means of reducing a discharge must have regard to cost. Here there can be more difficulty. In practice a survey of the particular industry as a whole can provide a guide to what is the best practicable means. In other words the best means of source control found by a survey of the industry can generally be taken as the provisional standard for all units in the industry on the argument that if one industrial unit can afford the means so can the others. As I have already said, however, this approach can be an over-simplification and some relaxation may have to be made in particular cases where need can be proven. Above all a reasonable time must be given to an industry to attain the standard. As experience is gained, a provisional standard can be made more stringent, applying at first to new plants and to modified plants.

Now that public concern strives for the maximum conservation of the environment, a more sophisticated approach is called for in setting source standards where the protection of health is not the primary criterion. First and foremost, there is a great need for cost benefit analysis of reducing waste discharges. The trouble is, however, that no reliable method of effectively applying cost benefit analysis appears to be available as yet. Without cost benefit analysis public pressure for unnecessarily high efficiency in reduction of waste discharges may build up on emotional grounds and be difficult to resist. It cannot be said too often in this context that the cost of reducing most discharges by 95% is not usually inordinate and the cost is not unduly disproportionate to the degree of reduction. Further reduction, however, very often escalates costs by more than a first order relationship, particularly when 99% and over is sought. The ultimate user of all industrial products is the public, at least when it comes to paying for them either directly or indirectly, and what is spent by the public in increased product prices because of pollution reduction cannot be spent on other desirable objectives in life, such as better education, better health services, better housing and so on. Cost benefit analysis could help in reaching rational decisions on priorities in these matters but until biological and sociological scientists can provide more detailed and quantitative assessments of pollution damage, the prospect of effectively applying this technique to the control of discharges do not appear very promising in the near future. Meantime those in authority over controlling pollution will have to continue making what are at times arbitrary decisions where the criteria of safety to health is not the critical factor. A purely emotional approach to problems of pollution must be kept well in check by all concerned with pollution.

2. Industrial Manufacturing Problems

So far this article has dealt with industrial pollution more from the viewpoint of its control. What are the technical problems of industry? Over the past 25 years in which I have worked as an official concerned with pollution control I have seen great strides forward by industry in controlling air pollution, more in the last decade than in the previous one, and I see no lack of technical knowledge to continue this progress. Industrial smoke from boilers and furnaces has been diminishing to the point where as a widespread problem it should become a thing of the past. Here industry has an incentive in that smoke from combustion represents loss of profit in wasted fuel. Even in processes such as baking carbon electrodes for dry cell batteries, the smoke of tarry emissions can be turned to use by burning to raise steam. Flares at oil refineries are a localised source of smoke but steam injection is doing much to reduce this problem and more could be done to hold and use waste gas. Coke ovens for making metallurgical coke as a batch operation are one of the few hard core problems in industrial smoke emission control. Until a continuous process of crabonising coal is adopted, I see no complete solution but it may well be that changes in the manufacture of steel will largely eliminate the need for metallurgical coke.

Apart from smoke, control of air pollution normally causes non-productive expenditure by industry. Emissions of particulate matter with a grain size around 1 μm and less, that is to say in the fume range of the spectrum of grain size in particles, are among the most expensive to control. Reduction of fume emission involves substantial running costs in the energy required for high pressure drops in arresters such as the venturi type of wet washer and bag filters, or high capital cost in electrostatic precipitation. In some manufacturing processes discharges of particulate matter as fume probably cannot be avoided because of the grain size of the product. On the other hand there are processes which chemical engineers could well apply their minds to avoid the formation of these particles within the production plants. I have in mind that some sulphuric plants can be operated without any significant mist in the discharges from their stacks whereas others have a chronic mist problem which can be solved only by non-productive expenditure on high efficiency filtration. An example of what can be done to prevent fume formation is a new fumeless technique for refining steel in electric arc furnaces. Government control of air pollution in Britain by the Alkali Act puts a limit of 0.12 gms/m^{-3} on the emission of the dense brown fume which arises when refining steel in electric arc furnaces is accelerated by injection of oxygen gas. High efficiency treatment plants are necessary to keep fume emission within the standard, and these plants involve substantial capital and running costs. Research, however, has led recently to a new way of accelerating the refining of steel in these furnaces by injection of crushed mill scale (iron oxide) by air which creates only a small amount of grey coloured fume which requires at most only simple and therefore inexpensive treatment to control emission of grit particles. Still a further example of preventing the formation of fume has been reported from the U.S.A. In the purification of molten aluminum recovered from scrap metal chlorine gas is commonly injected into the melt and this injection gives rise to the emission of a fume of chlorides which requires expensive high efficiency plants to arrest and the plant suffers from corrosion. The new process, of which details have not yet been released, is said to eliminate the chloride fume emission. No doubt other examples exist and there is certainly scope for chemical engineers to develop more processes of this kind which eliminate or reduce the need for treatment of waste fumes and so reduce non-productive expenditure.

Emission of waste gases normally present little problem nowadays to control by high efficiency absorption in wash liquors. Among recent developments, gas absorption by the floating ball type of absorber is perhaps worth a particular mention because it can in some cases reduce fume emissions as well as absorb gases. This feature can be of a special value in manufacturing processes where waste fumes and gases occur together. Water is commonly used as an absorption medium in treating waste gases and fumes but water is becoming scarce and expensive in some countries. Again the disposal of waste water can pose a pollution problem in its disposal. Here there is scope for chemical engineers to design waste gas treatments which make the most effective use of water and so reduce the overall costs of treatment. In some cases the stripping of absorbed gases for re-use may be the answer. Alternatively consideration might be given to the possibility of treating or concentrating waste waters to produce useful by-products. Water is not always the best choice of absorption medium for waste gases. Weak acid, for example, can sometimes be used to absorb fumes arising in the manufacture of the acid, and the weak acid can then be recycled into the process. In the case of hydrogen sulphide, organic liquids can be used and re-covered, and likewise in the case of some organic vapours.

3. Economic Views

Unless waste gases and fumes can be used in some way, waste treatment is a wholly non-productive expenditure, not only in absorbing the gases and fumes but also in treating the absorption media before disposal. In designing a manufacturing process it is no longer good enough for a chemical engineer to put arrows on his flow sheet indicating points and amounts of waste disposal for consideration at a later date what, if any, treatment is to be given to the wastes. He has to cost the treatment of all wastes and at the end of the day he may have to decide whether the cost of treatment is such that another manufacturing process might be more profitable. To minimise the overall cost of waste treatment a balance has to be struck between capital costs and running costs. Reliability of treatment plants has also to be considered closely, particularly if pollution control authorities insist on closure of the manufacturing process while repairs are carried out on treatment plants. Duplication of vital equipment such as pumps and fans may be necessary. Waste treatment plant of the lowest capital cost may not necessarily be the most economic in the long run when overall costs of maintenance and down time are reckoned. In practice many variables require to be optimised when designing a waste treatment plant for a particular purpose and the practice is growing of using computers to assist in making a design choice. Again computation of chimney dispersal of residues from waste treatment involves many variables which can be comprehensively handled only by a computer. Sometimes, however, local topography is such that mathematical computation of dispersal would not be reliable and in these cases resort has to be made to testing models in wind tunnels. In view of the very high cost of erecting tall chimneys, the cost of this detailed design work is relatively small and it can be very worthwhile.

4. Conclusion

Solutions to the problems of industrial air pollution lie in the hands of chemical engineers but the actual control of air pollution requires a multi-disciplinary approach involving scientists, technologists, economists and legislators. At the moment chemical engineers are well provided with all the techniques likely to be required of them but other disciplines have some way to go before they can provide definitive scientific targets. In the meantime chemical engineers could well take as their guide the key requirement of the century old British Alkali Act, namely, take the best practicable means to prevent the emission of noxious or offensive gases to the atmosphere and to render them harmless and inoffensive when discharged. This old requirement still poses a challenge to industry of today.

AIR POLLUTION CONTROL IN THE SWEDISH CHEMICAL INDUSTRY

GÖRAN A. PERSSON

Swedish National Environment Protection Board, Solna, Sweden

Abstract. Legislation, emission standards, and supervision are described are background, and industrial examples from Sweden are given with values of pollution from existing plants and standards for new ones.

The chemical industry is here defined as operations for the manufacture of inorganic and organic chemicals, nonferrous metals, ferro-alloys and cement.

The chemical industry is expanding very rapidly – 7% annual increase in production – resulting in a diversity of emission, e.g. of many different organic and metallic compounds from the plastics and petrochemical industries. Air pollution control in chemical industry will therefore require open eyes for new problems. These should be dealt with already in the process design. This paper gives the general background for air pollution control in Sweden and deals with problems in a number of installations within the chemical industry sector.

1. Legislation Background

From 1 July 1969 air pollution from stationary sources has been regulated by the Environment Protection Act. One distinguishing characteristic in the Act is prevention: the risk of potential harmful effects should be taken into account. Even though no proof of damage may exist, the harm of air pollution just be limited to a minimum by use of the 'best practicable means'.

A second characteristic of the Environment Protection Act is its concern for the total impact of a plant on the environment. A license is required for new installations and for the enlargement of existing installations; such control is necessary for reasons of potential air, water, and noise pollution. Registration is required, however, for installations with less potential for pollution. Finally, all potential sources of pollution can be forced to apply for a license.

The registration system is handled at the regional level by the county administrations, and the permit system at the central or regional level depending on the complexity of the pollution source. General control requirements for the Nation are established by the National Environment Protection Board.

In the field of air pollution the Board has adopted uniform emission standards for the entire country. The word standard is defined as a "limit adopted by a responsible agency". In the Swedish system, however, emission standards are not compulsory in themselves. Only the conditions laid down in the individual license have legal force. In this way the industries know in advance the general requirements for the prevention of pollution; yet at the same time, provisions can be made for industrial variances, if they are justified. This arrangement simplifies and accelerates the licensing.

2. Emission Standards

The Swedish emission standards described in this paper, if not otherwise stated, refer to the monthly arithmetical average of the total emissions. The standard preferably used is the "mass emission per process weight" regulation, E_m, in kg pollutant/ton product.

$$E_m = \frac{\sum\limits_{t=0}^{1 \text{ month}} m_t p_t}{\sum\limits_{t=0}^{1 \text{ month}} p_t} \qquad (m_t = \text{mass emission per process weight at time } t)$$

$(p_t = \text{process weight rate at time } t)$

The 'concentration' standard, E_c, in mg m^{-3} STP (dry gas), is also used.

$$E_c = \frac{\sum\limits_{t=0}^{1 \text{ month}} c_t n_t}{\sum\limits_{t=0}^{1 \text{ month}} n_t} \qquad (c_t = \text{concentration at time } t)$$

$(n_t = \text{stack gas volume rate at time } t)$

One of the more interesting results from the studies of emissions from existing installations was that the emissions during 'abnormal' conditions were found to be of considerable importance not only for the air quality in the vicinity during the episodes but also for the annual emissions. It is therefore quite clear that an effective control program must provide for the episodes also. In many industrial operations priority should be given to reliability and maintenance improvements instead of to the improvement of efficiency during favorable conditions.

The emission standards in Sweden are applicable to all operating conditions should be fulfilled during the intire life of the plant. This means that the control equipment must be dimensioned for emissions that are considerably lower than the numerical value of the standard. Generally, the equipment will also have to be divided into at least two independent units to avoid excessive emissions when one unit is not of operation.

3. Supervision

The supervision system is still in the development stage. It will include initial and periodic inspection in accordance with a detailed program to be worked out by the Environment Protection Board in collaboration with the industrial branch associations.

Within 3 to 6 months after a new or altered installation becomes operable the initial inspection will be carried out to insure compliance with the license. The inspections and emission measurements are mainly the responsibility of the county administrations, which are utilizing special authorized laboratoies for the source testing. The costs are met by the owner of the plant.

A very important part of the enforcement of emission standards is the continuing check on the emissions by the plant personnel. In the license it is normally specified that regular source tests should be made and that continuous recording instruments should be used as far as possible.

Emission data and operating data of interest should be available in the plant and forwarded at regular intervals to the county administration in accordance with a fixed program.

4. Inorganic Chemical Industry

The production of 'heavy' inorganic chemicals, existing emissions, and adopted emission standards are given in Tables I and II.

The emission standards for new sulfuric acid plants require the use of the 'double contact' process and demisters.

TABLE I

Data on production for 1970 and emissions at plants manufacturing inorganic chemicals in Sweden

Product (dominating process)	Number of plants	Range of production, 10^3 tons	Total production, 10^3 tons	Pollutant	Emissions, kg/ton product
Surfuric acid (Contact process)	9	4–175	700	SO_2	5–25
Hydrochloric acid (Mannheim process)	7	0.7–140	190	HCl SO_2	0.7–1.0 2.0–2.6
Phosphoric acid (Wet method)	4	1.5–50	70	F	0.13
Ammonia (Oil gasification)	4	6–65	140	SO_2	20–56
Nitric acid (Oxidation)	5	30–125	270	NO_x[a]	12–25
Chlorine, sodium hydroxide by electrolysis in mercury cells	8	4–75	322, 378	Hg	0.0015–0.01

[a] Calculated as NO_2

TABLE II

Emission standards for inorganic chemical industry in Sweden

Type of plant	Pollutant	Emissions, kg/ton product New units	Existing units
Sulfuric acid (Sulfur or pyrite as raw material)	Sulfur dioxide Sulfur trioxide	5 0.5	20 0.8
Chlorine	Mercury	0.001 (in ventilation air) 0.0005 (in hydrogen	0.005 0.001
Ammonia	Sulfur dioxide	Equipment for recovering sulfur	

Careful design is necessary to apply the standards for mercury emissions at chlorine manufacturing plants. Even if the emission of mercury to air is not considered to be a danger of the same order as emissions to water, stringent limits are necessary and justified. Investigations of mercury in fish have shown that metallic mercury may be converted by biological processes in nature to poisonous organic compounds.

4.1. FERROALLOY

The six ferroalloy plants with a total production of less than 200000 tons/year emit about 10000 tons of particulate matter annually. The inadequacy of available control technology is readily apparent from the adopted emission standards (Table III). These standards should be considered as provisional.

TABLE III

Ferroalloy emission standards for particulates in Sweden

Type of furnace	Emission standard, kg/ton product	
	New units	Existing units
Ferrosilicon	10	15
Ferrosilicon chromium	15	20
Ferrosilicon manganese	0.3	
Ferromolybdenum	3	3
Ferromaganese affiné	5	5

The emission of hydrogen fluoride from furnaces for manufacturing ferromolybdenum is limited to 1 kg HF/ton product.

To realize the control program for particulates, there is a need for investments of about 100 Sw. crowns/ton. Production costs after installation of control equipment are estimated to increase 30 Sw. crowns/ton ($6/ton).

4.2. CEMENT

The production of cement in 1968 was 4 million tons at seven plants. The annual

TABLE IV

Emission standards for particulates from cement plants in Sweden

Source	Emission standards, mg m^{-3} STP (dry gas)		Remarks
	New units	Existing units	
Kiln	250	500	Electrical precipitators at new kilns
Crusting, grinding, blending, and loading operations	250	250	should have at least two independent sections. The maximum emission with one section out of operation is 500 mg m^{-3} STP (dry gas)

emissions of particulates from cement manufacturing were estimated to be 10000 tons. The amount of gas treated per ton of cement is 5000 to 10000 m³ (standard temperature and pressure). The emission standards are given in Table IV.

In general the point sources in the cement industry are satisfactorily controlled. At the wet process 250 mg m⁻³ STP (dry gas) corresponds to about 215 mg m⁻³ STP and 140 mg m⁻³ at 150 °C. The main problem today is the control of ground sources. The cost of control will therefore be higher than that cost caused by the emission standards.

4.3. ALUMINIUM

Fluoride emissions from primary aluminium smelting are causing severe air pollution problems. In Sweden there is only one installation with a production capacity of 83000 tons/year. In the year 2000 about 600000 tons of aluminium may be produced in 3–4 installations. The present emissions from the Swedish installation are given in Table V.

TABLE V

Emissions from the Swedish installation for primary aluminium smelting

	Production tons Al/y	Water soluble fluorides kg/h	kg/ton Al	Total fluorides kg/h	kg/ton Al
Unit 1	13000	8.9	5.9	10.6	7.1
Unit 2	70000	6.4	0.8	8.7	1.1
Total	83000	15.3	1.6	19.3	2.0

They will be decreased by 50% within two years in connection with a planned expansion of 50000 tons/year.

5. Air Quality Requirements

A common air pollutant in the chemical industry is sulfur dioxide, which is emitted from processes and from fuel combustion. The air quality standard for sulfur dioxide to prevent health effects is shown in Table VI. However, sulfur dioxide is also causing

TABLE VI

Air quality standards for sulfur idoxide in Sweden

Time average	Concentration, ug m⁻³	ppm	Remarks
1 month	143	0.05	Should not be exceeded
24 h	286	0.10	Should not be exceeded more than once per month
30 min.	715	0.25	Should not be exceeded more than 15 times per month (1 % of the time)

Danckwerts the fraction of the surface which is replaced in unit time. The difference between the theories of Higbie and Danckwerts is that the former assumes that all surface elements remain equally long at the surface whereas the latter assumes that the probability for replacement is independent of the time the element has stayed at the surface. When the three models are applied to practical cases involving mass-transfer with chemical reaction, the results differ very little numerically, although the mathematical functions appearing in the solutions can be of different types. It has not been possible to devise experiments which will show conclusively which description comes closest to reality. It is now generally accepted that all three theories are applicable, and that the choise in each case is dictated by mathematical convenience. As the Whitman film theory is much easier to visualize, it will be used almost exclusively in this paper. The treatment of gas liquid reactions consists essentially of combining these models with the kinetics of the chemical reaction taking place.

It is important to note that both the film theory and the surface renewal theories presuppose that the diffusion region is small compared to the thickness of the under-lying liquid layer. For liquids flowing over packings the average thickness of the liquid layer is equal to (l/a) or the ratio of the hold up l to the surface of the packing a, both per unit volume of the column. The thickness of the diffusion film is

$$\delta = D_A/k_L, \tag{1}$$

or the ratio of the diffusivity to the liquid side mass transfer coefficient. The ratio of the two becomes:

$$\frac{\text{Diffusion film}}{\text{liquid layer}} = \frac{D_A a}{k_L l}. \tag{2}$$

For the $\frac{1}{2}''$ to the $1\frac{1}{2}''$ Raschig rings this ratio usually varies between 10^{-1} and 10^{-3} indicating that the condition for the applicability of the models usually is fulfilled. It must be born in mind, however, that in a packed column the thickness of the liquid layer is not uniform, but varies greatly from place to place. We shall return to this point later.

2. General Considerations

It is useful to distinguish between the case where chemical reaction and molecular diffusion occur simultaneously and where they occur consecutively. In the latter case the chemical reaction is so slow that it does not take place to any significant extent in the diffusion film, but only in the bulk of the liquid, and even there perhaps only to a limited extent. In this case the equations appropriate to purely physical mass transfer can be applied to the diffusion film, the chemical kinetics giving the back-pressure of the diffusing species in the bulk of the liquid, and hence the driving force for the mass transfer. The conditions for negligible reaction in the diffusion film can easily be derived. It is convenient to use the film concept, but the conclusions are valid also for the surface renewal theories. The film thickness is given by Equation (1). Assuming first order kinetics, the rate of reaction is directly proportional to the concentration:

$$r = k_1 A.$$

The concentration of the diffusing species can in no case be larger than the physical equilibrium value, i.e. $A < A^*$. The condition for the amount reacted in the diffusion film to be small compared to the amount transported through the film, can then be expressed as follows:

$$\frac{D_A}{k_L} k_1 A^* \ll k_L (A^* - A^0). \tag{3}$$

With complete reaction in the bulk of the liquid $A^0 = 0$ the condition becomes:

$$\frac{D_A k_1}{k_L^2} \ll 1. \tag{4}$$

The corresponding expression for second order reactions

$$A + B \rightarrow \text{products with } r = k_2 AB, \tag{5}$$

is obtained simply by replacing k_1 by $k_2 B^0$:

$$\frac{D_A k_2 B^0}{k_L^2} \ll 1 \tag{6}$$

For several applications it is important to know also the condition for complete reaction in the bulk of the liquid i.e. $A^0 \ll A^*$. The rate of transfer through the diffussion film must in this case be equal to the rate of reaction in the bulk of the liquid:

$$a k_L (A^* - A^0) = k_2 A^0 B^0 l,$$

or since $A^0 \ll A^*$:

$$\frac{A^0}{A^*} \approx \frac{a k_L}{l k_2 B^0} \ll 1. \tag{7}$$

We now have at our disposal criteria for negligible chemical reaction in the diffusion film Equation (6) and for complete reaction in the liquid bulk Equation (7). A particularly interesting situation arises when both these criteria are satisfied, because that would point the way to useful methods for determining $k_L a$. Rearranging Equations (6) and (7) gives:

Thickness of diffusion film:

$$\delta = \frac{D_A}{k_L} \ll \frac{k_L}{k_2 B^0}, \tag{8}$$

Total thickness of liquid layer:

$$\frac{l}{a} \gg \frac{k_L}{k_2 B^0}. \tag{9}$$

Taken together Equations (8) and (9) give the obvious and necessary requirement for the fulfillment of both criteria that the thickness of the diffusion film must be small compared to the total thickness of the liquid layer. Equally obvious, this requirement is not sufficient. In addition it is necessary to select chemical conditions such that k_L/k_2B^0 has a value that falls nicely between the thickness of the diffusion film and the total liquid layer. In some cases this can be done by using catalytic reactions and adjusting the amount of catalyst. It is pertinent to point out that in using absorption of oxygen in sulfite with cobalt ions as catalyst, sufficient care has not always been taken to make sure that Equations (6) and (7) are satisfied. Incidentally, the reaction is not first order with regard to oxygen, but second order.

3. The Diffusion Equations

The mathematical treatment of mass transfer with chemical reaction is based on two fundamental equations. The first is the diffusion equation

$$F = - D \frac{\partial c}{\partial x}, \tag{10}$$

and the second a material balance equation

$$D \frac{\partial^2 c}{\partial x^2} = \frac{\partial c}{\partial t} + r. \tag{11}$$

Equation (11) shows that if at any point the rate of reaction is infinite $\partial^2 c/\partial x^2$ becomes infinite indicating a discontinuity in the concentration gradient $c(x)$. With a knowledge of the kinetics of the reactions i.e. the rate of reactions expressed as functions of the temperature and the concentrations of the relevant molecular species, the material balance equations can be solved using the proper boundary conditions. As a result the concentrations appear as functions of distance from the surface. For non-steady state situations the concentrations are of course also functions of time. Generally, we are not interested in the concentration gradients, but only in the amount of gas absorbed. This can be found by applying the diffusion equation to the situation at the surface:

$$R = - D_A \left(\frac{\partial c}{\partial x} \right)_{x=0}. \tag{12}$$

The steady state equation with $\partial c/\partial t = 0$ applies to the film theory whereas the surface renewal theories require the unsteady state equation. This, however, does not mean that the mathematics of the surface renewal theories are more difficult to solve. In some cases it is actually the other way round.

4. Chemical Reactions in the Diffusion Film

The effect of chemical reactions on the rate of mass-transfer is usually expressed as an enhancement factor E defined as the ratio of the rate of absorption with chemical reaction to the rate of purely physical absorption in the absence of chemical reaction:

$$\bar{R} = Ek_L(A^* - A^0).$$ (13)

The enhancement factor is always greater than 1.

We shall now consider the case with an irreversible second order chemical reaction

$$A + B \rightarrow \text{products, with } r = k_2 AB.$$

A is here the species originally present in the gaseous phase whereas B is the second reactant present exclusively in the liquid phase. It is assumed here that the reaction is fast enough to take place exclusively in the diffusion film.

It is convenient to distinguish two cases according to whether the chemical reaction is infinitely fast or not, that is infinitely fast compared to the rate of transport of the solute A through the diffusion film. We shall first consider absorption with infinetely fast irreversible chemical reaction, and in so doing lean on the pioneering work of Hatta (1928).

It is obvious that with an infinitely fast irreversible reaction both molecular species A and B cannot be present together anywhere in the diffusion film. The corresponding diffusion gradients are illustrated in Figure 1. The effect of the chemical reaction is to reduce the thickness of the diffusion film for the component being absorbed. The corresponding enhancement factor can easily be shown to be

$$E = \frac{\bar{R}}{k_L A^*} = \left(1 + \frac{D_B B^0}{D_A A^*}\right).$$ (14)

The enhancement factor thus does not depend on the rate of the chemical reaction, provided the rate is great enough. A criterion is clearly required for deciding whether a given reaction can be treated as instantaneous. This will be discussed later.

In the general case of second order irreversible reactions the reaction will not be limited to a narrow reaction zone, but take place throughout the diffusion film. Using the film theory, the partial differential Equation (11) becomes an ordinary differential equation, and applied to the two diffusing molecular species A and B (Figure 2):

$$D_A \frac{d^2 A}{dx^2} - k_2 AB = 0,$$
$$D_B \frac{d^2 B}{dx^2} - k_2 AB = 0,$$ (15)

with the boundary conditions

$$\text{for } x = 0: \quad A = A^*,$$
$$\text{for } x = \delta: \quad A = 0 \quad \text{and} \quad B = B^0.$$

It is illuminating to render these equations dimensionless by introducing the following

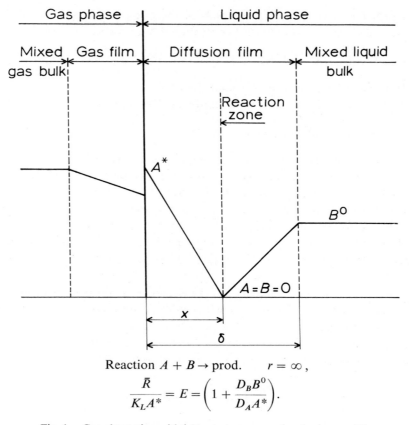

Reaction $A + B \rightarrow$ prod. $r = \infty$,

$$\frac{\bar{R}}{K_L A^*} = E = \left(1 + \frac{D_B B^0}{D_A A^*}\right).$$

Fig. 1.   Gas absorption with instantaneous second order irreversible reaction (according to Hatta, 1932).

dimensionless variables:

$$x' = \frac{x}{\delta} = \frac{x k_L}{D_A}; \qquad A' = \frac{A}{A^*} \quad \text{and} \quad B' = \frac{B}{B^0}. \tag{16}$$

After rearranging Equations (15) become

$$\frac{d^2 A'}{dx'^2} = \frac{D_A k_2 B^0}{k_L^2} A'B' = 0,$$

$$\frac{D_B B^0}{D_A A^*} \frac{d^2 B'}{dx'^2} - \frac{D_A k_2 B^0}{k_L^2} A'B' = 0. \tag{17}$$

Introducing the dimensionless groups

$$M = \frac{\sqrt{D_A k_2 B^0}}{k_L} \quad \text{and} \quad m = \frac{D_B B^0}{D_A A^*}. \tag{18}$$

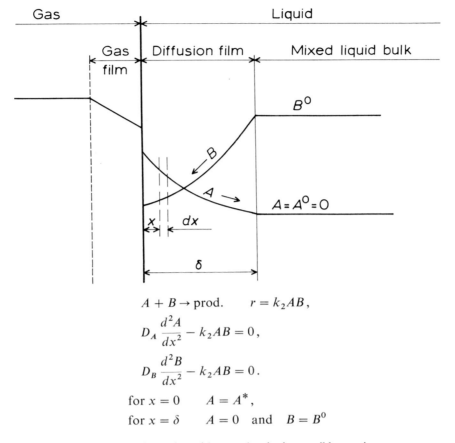

$$A + B \rightarrow \text{prod}. \qquad r = k_2 AB,$$

$$D_A \frac{d^2 A}{dx^2} - k_2 AB = 0,$$

$$D_B \frac{d^2 B}{dx^2} - k_2 AB = 0.$$

for $x = 0$ $A = A^*$,

for $x = \delta$ $A = 0$ and $B = B^0$

Fig. 2. Gas absorption with second order irreversible reaction.

Equation (17) becomes:

$$\frac{d^2 A'}{dx'^2} - M^2 A'B' = 0,$$

$$m \frac{d^2 B'}{dx'^2} - M^2 A'B' = 0,$$

(19)

with the boundary conditions

for $x' = 0$: $A' = 1$,

for $x' = 1$: $A' = 0$ and $B' = 1$.

(20)

Remembering that the rate of absorption is given by Equation (12):

$$\bar{R} = - D_A \left(\frac{dA}{dx} \right)_{x=0},$$

the enhancement factor becomes in dimensionless terms

$$E = \frac{\bar{R}}{k_L A^*} = -\left(\frac{dA'}{dx'}\right)_{x'=0}. \tag{21}$$

These considerations show that the enhancement factor must be a function of two and only two dimensionless parameters:

$$E = E(M, m). \tag{22}$$

The pair of simultaneous differential equations, Equation (19) cannot be solved to give an accurate analytical solution, but numerical solutions and an approximate solution are given by van Krevelen and Hoftijzer (1948) and reproduced in Figure 3. This figure also covers the special cases with instantaneous reaction and with pseudo first order reaction. The criteria for the different types are derived from the figure itself and

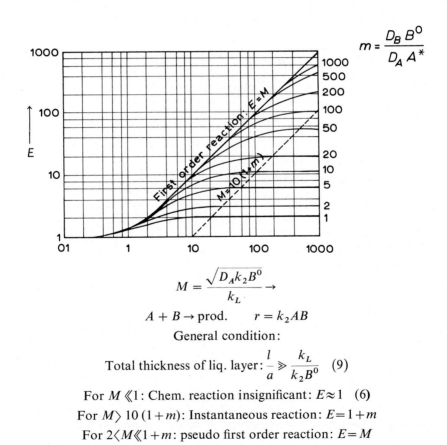

$$M = \frac{\sqrt{D_A k_2 B^0}}{k_L} \rightarrow$$

$$A + B \rightarrow \text{prod.} \qquad r = k_2 AB$$

General condition:

Total thickness of liq. layer: $\dfrac{l}{a} \gg \dfrac{k_L}{k_2 B^0}$ (9)

For $M \ll 1$: Chem. reaction insignificant: $E \approx 1$ (6)

For $M > 10(1+m)$: Instantaneous reaction: $E = 1 + m$

For $2 < M \ll 1 + m$: pseudo first order reaction: $E = M$

Fig. 3.   Enhancement factors for second order irreversible reactions
(according to Van Krevelen, 1955).

given underneath. Note that the criterion for insignificant chemical reaction is identical with that derived earlier and given as Equation (6).

Absorption of substances which dissociate into ions on entering the liquid phase is in principle a case of gas absorption with chemical reaction. However, if the dissociation is instantaneous, as in the case of hydrogen chloride, and no other reaction takes place, it can still be treated as pure physical absorption. Although the positive hydrogen ion intrinsically has a greater diffusivity than the negative chloride ion, the principle of electroneutrality requires the two ions to move together, the fast moving hydrogen ion speeding up the slow moving chloride ion and the slow moving chloride ion slowing down the hydrogen ion. As a result, an average diffusion coefficient can be used.

This simplified treatment may not be justified when chemical reactions take place in the diffusion film. Sherwood and Wei (1955) have illustrated the effect of taking the different ionic diffusivities into account. For instance, they have considered the absorption of HCl by a solution of NaOH. This reaction can be regarded as instantaneous and irreversible:

$$H^+ + OH^- \rightarrow H_2O.$$

The sodium and chloride ions do not take part in the reaction. The distribution of the various species of ions in the film is shown in Figure 4. In this case the fast moving hydrogen- and hydroxyl-ions can move ahead of their slow moving partners, the chloride- and sodium-ions. It is noteworthy that in the reaction zone the principle of

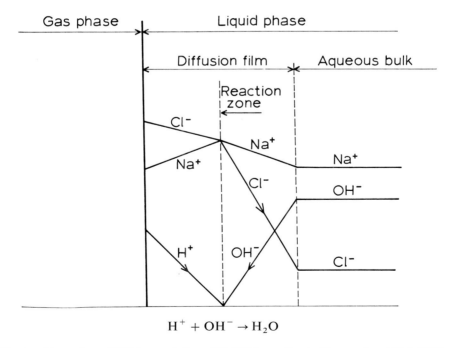

$$H^+ + OH^- \rightarrow H_2O$$

Fig. 4.   Absorption of HCl by a solution of NaOH (according to Sherwood and Wei, 1955).

electroneutrality requires the concentrations of sodium and chloride ions to be equal. It is not easy to grasp the significance of these effects by verbal arguments, but the results of the calculations illustrated in Figure 5 show that enhancement factors calculated by considering the ions individually, the upper lines, may be more than twice those calculated by the simple Hatta theory using molecular diffusivities, the bottom

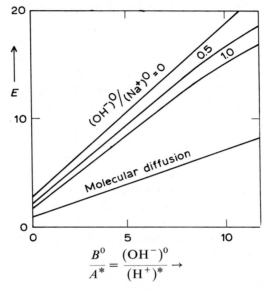

Fig. 5.   Enhancement factors for absorption of HCl by a solution of NaOH
(according to Sherwood and Wei, 1955).

line marked molecular diffusivity. The effect is greatest when sodium chloride is present in large excess, as shown by the line marked $[OH^-]^0/[Na^+]^0 = 0$.

5. Chemical Methods for Measuring Interfacial Area and Mass-Transfer Coefficients

The quantity of solute transferred from the gaseous to the liquid phase per unit volume of a contactor depends on the product of the liquid side mass-transfer coefficient K_L and the interfacial area per unit volume a:

$$dQ = K_L a (A^* - A^0) \, dV. \tag{23}$$

The value of $k_L a$ can be determined by purely physical measurements, but two serious difficulties are encountered: one is that in an efficient contacting device such as a sieve plate, the two phases may approach equilibrium very closely and extremely accurate determinations of flows and concentrations become necessary. The other difficulty is that the flow pattern and the residence time distribution of the phases may be complicated and unknown, so that the value of A^0 as a function of A^* cannot be determined and Equation (23) cannot be integrated.

The chemical methods of determining $k_L a$ avoid these difficulties. In the ideal situation the chemical reaction is slow enough not to take place in the diffusion film, but fast enough to be completed in the bulk of the liquid. The necessary criteria are embodied in Equations (8) and (9). A convenient chemical system is the absorption of carbondioxide into a bicarbonate buffer solution. At $pH > 10$ the predominating reaction is:

$$CO_2 + OH^- \rightarrow HCO_3^-, \tag{24}$$

which is first order with respect to carbondioxide and hydroxyl ions:

$$r = k_{OH^-} [CO_2] [OH^-]. \tag{25}$$

The concentration of OH^- is determined by the ratio of carbonate ions to bicarbonate and the product $k_{OH^-} - [OH^-]$ can consequently be varied in the range 0.1 to 3.0 s^{-1}. The conditions (8) and (9) can therefore in most cases be satisfied. Another well-known system is the absorption of oxygen into sulfite solutions containing cobalt or copper ions as catalysts. This reaction appears to be second order with respect to oxygen and zero order to sulfite. The rate is, however, strongly dependent on pH.

In case it is not possible to satisfy the condition that the reaction shall go to completion in the bulk of the liquid, it is still possible to determine $k_L a$, but it is now necessary to carry out a series of experiments with different concentrations of the second component.

There is another quite different method whereby a chemically reacting system can be used to determine $k_L a$. The reaction must first of all satisfy the criterion for instantaneous reaction given in Figure 3. If, furthermore, the conditions are selected so, that the concentration of the reactant in the liquid B^0 is much greater than the solute equilibrium concentration A^*, it is evident from Equation (14) that

$$\bar{R}a = k_L a B^0 \left(\frac{D_B}{D_A} \right). \tag{26}$$

Under these conditions the rate of mass transfer is independant of the equilibrium concentration of the solute being transferred. It is an obvious further condition that the gas film resistance must be negligible. The condition for this is easily derived:

$$k_G p_A \gg k_L B^0. \tag{27}$$

Suitable chemical systems are the absorption of ammonia in sulphuric acid or of sulphur dioxide, chlorine or hydrogen chloride in solutions of alkalis. Reference is made to Sharma and Danckwerts (1970).

In order to determine the interfacial area fairly slow, pseudo-first order reactions can be used. The conditions for this can be seen from Figure 3. In this case

WEATHER AND CLIMATE FACTORS IN INDUSTRIAL SITE EVALUATION WITH RESPECT TO AIR POLLUTION

NIELS E. BUSCH

Danish Atomic Energy Commission, Research Establishment Risø, Roskilde, Denmark

Abstract. The paper gives a brief introduction to air pollution meteorology with particular emphasis on the description of short-range dispersion from single sources. Industrial sites are normally not chosen on the basis of air pollution criteria. However, air pollution problems may be avoided to some extent, if meteorological, climatological, topographic, and aerodynamic factors are taken into account in the design and location of industrial stacks.

A long series of papers reviewing various aspects of air pollution meteorology and climatology has been published during the last few years (Slade, 1968; Stern, 1968; Panofsky, 1969; Forsdyke, 1970; McCormick, 1970; Pasquill, 1971; Pack, 1971). In this short article I cannot possibly compete with the three hundred pages which Slade (1968) uses to explain the fundamentals, nor can I hope to achieve what takes the consecutive efforts of Wanta, Strom, McCormick, Hilst and Hewson plus two hundred and twenty-two pages of Stern's Air Pollution to accomplish. Thus those who are already familiar with the theme are unlikely to learn much that is new to them.

On the other hand those who know nothing about the subject, but who want to familiarize themselves with its practical aspects, probably will be disappointed, too. The reason is that although modern techniques have propelled meteorology, and its subdiscipline micrometeorology, out of the witchcraft era, it is still – owing to the complex nature of even the simplest problems – a science which relies heavily on qualitative and phenomenological arguments. Therefore any introductory lecture will have to be either wrong or confusing.

1. Site Evaluation

Sites for industrial plants as well as nuclear reactors are not chosen solely – or to any significant extent – on the basis of air pollution criteria. Availability of power, manpower, cooling water and other resources, quality of roads and other means of transportation, distance to customers, building and labor costs and many other factors are considered in the decision making. Furthermore most industries are where they are and cannot be moved around as the wind blows. Hence the task at hand is to evaluate specific geographical locations in which pollution sources of a specified type either exist, are planned, or may come into existence accidentally.

There is of course in principle no difference between routine releases of pollutants and releases due to accidents. In practice, however, the latter will be of short duration and little interest unless the material released is toxic or otherwise a threat or major nuisance.

If an expected release is evaluated to yield unacceptable concentrations under some

G. Lindner and K. Nyberg (eds.), Environmental Engineering, 81–94. All Rights Reserved

external conditions, then there are two possibilities:

(1) Reduction of emission by either reduction of load, installation of better filters, change of process or storage of pollutants until release can be carried out safely

(2) Enhancement of dispersion by either redesign of source (e.g. increase of effective stack height) or relocation.

Determination of what is a reasonable or prudent commitment to site evaluation depends on a series of ecological, political, economic and technical factors which we cannot discuss here. It is clear, however, that the basic questions are: What are the consequences of the release and how much will it cost to design the plant to be on the safe side?

2. The Effects of Air Pollution

If we assume that the amount of a pollutant which a member of a population absorbs is proportional to the concentration of the pollutant to which he is exposed, we may write

$$D(t, \tau) = K_e \int_{t-\tau}^{t} C(t') \, dt', \tag{2.1}$$

where D is the dose inhaled during the time period τ, $C(t)$ is the concentration as a function of time t, and K_e is a factor describing the efficiency with which the pollutant is absorbed.

The effect of a given dose depends not only on its magnitude. Also the rate at which the dose is absorbed may be important. Figure 1 illustrates schematically the effect on

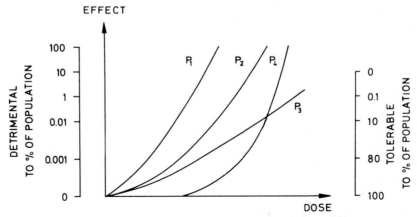

Fig. 1.   Effect of dose for a fixed dose time (schematic).

a population of four hypothetical pollutants for a fixed dose time τ. Figure 2 shows crudely how the effect of a given dose may behave as a function of dose time.

The simplest and still in some cases realistic assumption we can make is that the

EFFECT

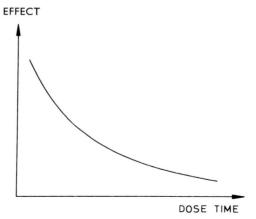

DOSE TIME

Fig. 2.   Effect of dose time for a given dose (schematic).

effect of a pollutant can be expressed

$$E = K(\tau, \bar{C}_\tau) \tau \bar{C}_\tau(t),$$

(2.2)

where the average concentration

$$\bar{C}_\tau(t) = \frac{1}{\tau} \int_{t-\tau}^{t} C(t') \, dt'.$$

(2.3)

The function $K(\tau, C)$ describes both the efficiency with which the member is absorbing the pollutant and his sensitivity to that particular pollutant.

We require that the probability of the effect – on a randomly picked member of the population – being bigger than some critical value E_c is smaller than some acceptable value P_c, i.e.,

$$\text{Prob}(E > E_c) < P_c,$$

(2.4)

or

$$\text{Prob}(K > K_c(\tau, C)) \, \text{Prob}(\bar{C}_\tau > C) < P_c,$$

(2.5)

where $\text{Prob}(\bar{C}_\tau > C)$ is the probability of the average concentration $\bar{C}_\tau$ exceeding a value C, and $\text{Prob}(K > K_c(\tau, C))$ is the probability of the sensitivity K being bigger than some critical value K_c. Both probabilities are extremely difficult to assess; we shall not discuss the difficulties involved in prescribing 'acceptable' values for P_c.

Anyhow, if for a given source and pollutant the meteorologists could provide the probability distribution for $\bar{C}_\tau$ for all values of τ, then they could leave it to somebody else to worry about the rest.

In reality the problems are of course likely to be considerably more complicated than this simple example indicates. In general the efficiency function $K(\tau, C)$ will be dependent on time t and the presence of other pollutants. Furthermore one would in

4. Atmospheric Motion on all Scales

Figure 4 depicts schematically the instantaneous picture of smoke coming out of stack into a turbulent atmosphere. It is clear that the motion transporting the smoke consists of a superposition of modes ('eddies') with different characteristic length scales (sizes). In the atmosphere there is motion on all scales from the smallest turbulent eddies with length scales of less than a millimetre to general circulation with length scales of the order of 10 000 kilometres (Figures 5 and 6).

Unfortunately there is no sharp distinction between the various types of atmospheric motion, but rather an ever changing continuous distribution of kinetic energy on a continuum of scales in space and time. Smoke out of a chimney will be dispersed by

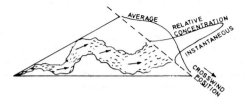

Fig. 4.   Instantaneous and average aspects of the cross-wind spread of a smoke-plume.

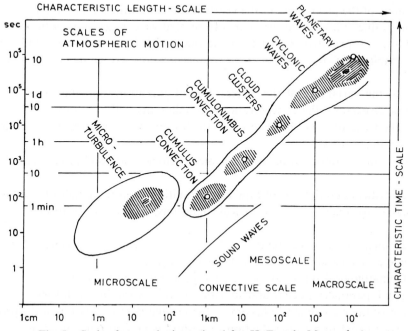

Fig. 5.   Scale of atmospheric motion (after H. Fortak, *Meteorologie*,
Deutscher Buchgemeinschaft Darmstadt, 1971).

the smallest eddies (i.e., eddies smaller than the cross wind dimensions of the plume) while the bigger eddies will bend, wiggle and advect the plume.

We shall not go deeper into the crucial problem of how to separate the dispersion from the advection (i.e., the turbulent diffusion from the transport by the mean wind) It is sufficient to mention that it appears reasonable to regard fluctuations with periods

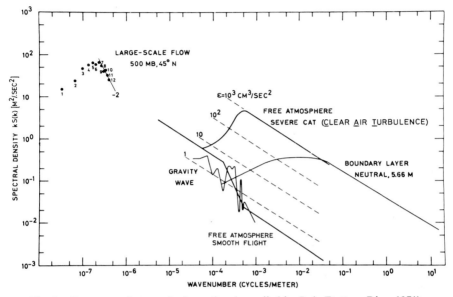

Fig. 6.   Spectrum of atmospheric motion (compiled by J. A. Dutton, Risø, 1971).

less than ten minutes to an hour as belonging to the turbulent regime, and those with time scales longer than approximately one hour as variations of the mean flow (Fiedler and Panofsky, 1970).

As the plume grows down-stream, bigger and bigger eddies become active in dispersing the plume. Hence the rate at which a plume is spreading is not only dependent on the intensity of the turbulent motion and the distribution of the intensity on eddy size, but also on the dimensions of the plume, i.e., on travel time from the source.

This is the reason why ordinary diffusion equations have not been applied too successfully to the problem of diffusion from stacks. The more successful approaches have been entirely empirical. A typical scheme (Pasquill, 1962; Slade, 1968; Fiedler, 1969) is described in the next section.

5. A Simple Model for Turbulent Dispersion

We assume that the flow is approximately stationary (in a statistical sense) over periods of the order of one hour or longer. This means that the model can be expected to hold for horizontal distances up to some tens of kilometres. We furthermore assume that the shape of the cross-wind distribution of effluent concentrations is known, when the concentrations are averaged over one hour.

It is usually assumed that the distribution is Gaussian, so that the concentration may be written

$$C = \frac{Q}{2\pi u \sigma_y \sigma_z} \exp\left\{-\frac{y^2}{2\sigma_y^2}\right\}\left[\exp\left\{-\frac{(z-H)^2}{2\sigma_z^2}\right\} + \exp\left\{-\frac{(z+H)^2}{2\sigma_z^2}\right\}\right],$$

(5.1)

where: Q is the source strength, u the mean wind speed at the effective stack height, y and z the horizontal and vertical distances from the center line of plume, respectively, σ_y and σ_z the lateral and vertical standard deviations of the concentration distribution, H the effective stack height.

The expression implies total reflection of effluent at the ground and conservation of mass.

Using existing empirical evidence we relate the standard deviations σ_y and σ_z to measurable meteorological quantities such as σ_θ and σ_α, i.e., the standard deviations of the vertical and horizontal wind directions, respectively, and to the distance from the source. Smith (1968) has suggested relations of the type

$$\sigma_y = b\sigma_\alpha x^\beta,$$
$$\sigma_z = b\sigma_\theta x^\beta,$$

(5.2)

where x is the distance down wind from the source and b and β are dependent on the density stratification of the atmosphere.

According to (5.1) the concentrations at ground level are

$$C_{z=0} = \frac{Q}{\pi u \sigma_y \sigma_z} \exp\left\{-\frac{y^2}{2\sigma_y^2}\right\}\exp\left\{-\frac{H^2}{2\sigma_z^2}\right\}.$$

(5.3)

Figure 7 (after Fiedler, 1969) shows a typical concentration pattern on the ground.

If climatological records of u, wind direction, σ_α, and σ_θ are available or if these quantities can be established some other way on an hourly basis, say, then the ground level concentrations can be computed on hourly basis for a given source strength and height. By use of such climatological material we can say where, how often, and for how long the hourly average of ground level concentrations exceeds a critical value. Figure 8 (after Fiedler, 1969) shows the frequency with which the relative concentra-

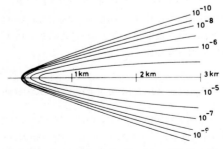

Fig. 7.   Relative concentrations at the ground (after Fiedler, 1969).

tion $(Q=1)$ according to the model should exceed 2×10^{-7} near Munich (50 m stack height). For comments on the effect of sampling time on measured concentrations see, for example, Hino (1968), Saltzman (1970), and McGuire and Noll (1971).

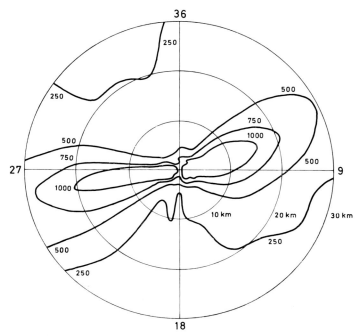

Fig. 8.   Frequency of relative ground level concentrations $\geqslant 2 \times 10^{-7}$ (after Fiedler, 1969).

6. Atmospheric Stability and Surface Roughness

The intensity of the turbulent motion as measured by σ_α and σ_θ in the lower atmosphere is highly variable. Two mechanisms are producing turbulence: (1) Aerodynamic instability of shear layers (mechanically produced turbulence) and (2) Heating from below (convective or thermal turbulence). In the planetary boundary layer the mechanical production is essentially equal to $K_M (\partial u/\partial z)^2$. The thermal production, which may be negative, is equal to $K_H(g/T)(-\partial\theta/\partial z)$, where $\partial\theta/\partial z$ is the vertical gradient of the mean potential air temperature, and K_M and K_H are the turbulent diffusivities for momentum and heat, respectively[*]. The negative ratio of the thermal to the mechan-

---

[*] The potential temperature of an air parcel is defined as the temperature the parcel would assume if brought isentropically to a standard pressure. In the atmospheric boundary layer we have to a good approximation (no phase transformations):

$$\theta = T + \Gamma z + \text{const.} \quad \text{or} \quad \partial\theta/\partial z = \partial T/\partial z + \Gamma$$

with $\Gamma = g/c_p$, where g is the acceleration due to gravity and c_p is the specific heat at constant pressure. The turbulent diffusivities are defined through

$$Q = -c_p\varrho K_H \frac{\partial\theta}{\partial z} \quad \text{and} \quad \tau = \varrho K_M \frac{\partial u}{\partial z}$$

where Q and τ are the vertical heat- and momentum fluxes, respectively; ϱ is the air density.

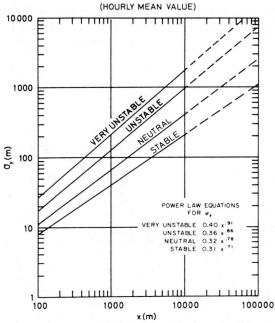

Fig. 9.   Horizontal standard deviations of a plume (after Smith, 1968).

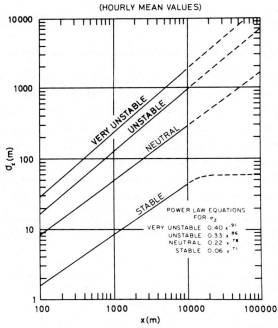

Fig. 10.   Vertical standard deviations of a plume (after Smith, 1968).

ical production is a nondimensional number called the flux Richardson number

$$\mathrm{Rf} = -\frac{\text{thermal prod.}}{\text{mech. prod.}} = \frac{K_H}{K_M}\frac{g}{T}\frac{(\partial T/dz + \Gamma)}{(\partial u/dz)^2} = \frac{K_H}{K_M}\mathrm{Ri},$$

where Ri is the gradient Richardson number, g the acceleration due to gravity, T the temperature, and $\Gamma \simeq 1\,^\circ\mathrm{C}/100$ m is the dry-adiabatic lapse rate. If $(\partial T/\partial z) < -\Gamma$, i.e., Ri < 0, then the atmosphere is said to be unstable and both mechanisms produce turbulence. In such situations the turbulence is intense and mixing proceeds rapidly. If $(\partial T/\partial z) > -\Gamma$, so that Ri > 0, then the thermal production term is negative and causes anihilation of turbulent motion. If Ri > 0 the atmosphere is said to be stably stratified. At a certain Richardson number, the anihilation of turbulence because of buoyancy damping and by viscous dissipation becomes too strong for turbulence to exist, the flow becomes laminar and the mixing very poor.

Figures 9 and 10 illustrate the drastic effect which the thermal stratification has on turbulent dispersion.

The roughness of the ground as well as its temperature, heat capacity and thermal conductivity play a role, too. The rougher the surface is the greater the mechanical production and therefore the better the mixing, all else being equal. In the lowest layers over densely urbanized areas, where the roughness is pronounced and the surface usually warmer than the air above it, stable stratification is rare, the mechanical production big, and the mixing is consequently thorough.

7. Effective Stack Height

When a hot plume is released into the atmosphere it will rise above the stack owing to its excess of buoyancy. As it mixes with the colder ambient air it loses buoyancy and the ascent rate decreases. Eventually the plume may level off at the effective stack height defined by $H = h + \Delta h$, where h is the height of stack and Δh the plume rise distance.

In unstable or neutral atmospheres the plume should continue to rise indefinitely or at least until it encounters an inversion. In practice, however, it mixes so rapidly with the ambient air that further plume rise can be neglected beyond a down wind distance of approximately 10 stacks heights.

In stable air the plume rises to a final height, but the distance it takes depends on the thermal stability.

Briggs (1969) advocates the following formulae for the bent-over buoyant plume (see also Hewett *et al.*, 1971).

Unstable and neutral atmosphere:

$$\Delta h = 1.6\,\frac{F^{1/3}}{u}\,x^{2/3} \quad \text{for} \quad x \lesssim 10h,$$

$$\Delta h = 1.6\,\frac{F^{1/3}}{u}\,(10h)^{2/3} \quad \text{for} \quad x \gtrsim 10h. \tag{7.1}$$

Stable atmosphere:

$$\Delta h = 1.6 \, (F^{1/3}/u) \, x^{2/3} \quad \text{for} \quad x \lesssim 2.4 \, us^{-1/2} ,$$
$$\Delta h = 2.9 \, (F/us)^{1/3} \quad \text{for} \quad x \gtrsim 2.4 \, us^{-1/2} . \tag{7.2}$$

Here F is a buoyancy parametre proportional to the heat emission Q_H through the stack:

$$F \simeq 3.8 \times 10^{-5} \, Q_H , \tag{7.3}$$

where F is in $m^4 \, s^{-3}$, if Q_H is in cal s^{-1}. The stability parameter s is proportional to the vertical gradient of potential temperature:

$$s = \frac{g}{T} \frac{\partial \theta}{\partial z} . \tag{7.4}$$

The plume rise can be considerable. If we assume that 5% of the load on a 200 MW power plant goes up the chimney, then the plume rise amounts to approximately 100 m at a distance of 500 m from the stack in neutral stratification with a 5 m s^{-1} wind.

One situation is of particular interest: Release of a buoyant plume in a calm or almost windless, stably stratified atmosphere. The effluent will rise (Briggs, 1969) above

$$\Delta h = 5F^{1/4}s^{-3/8} , \tag{7.5}$$

the stack and form a blini (a Russian pancake) or a ribbon. When the inversion breaks up from below the pollutants will get washed down to the ground and give rise to relatively high surface concentrations of short duration (30–45 min; Carpenter et al., 1971). In a calm, isothermal atmosphere the example used above yields $\Delta h \simeq 300$ m.

8. Complications

Although the simple dispersion model discussed above appears involved enough to the practioner, it will in many situations be far from realistic enough. The problem quickly becomes too difficult to handle, if we are not studying turbulent dispersion in stationary, homogeneous turbulence – and we never are in practice. Some of the most usual complications shall be mentioned briefly. Although models exist for most of them it is fair to say that the importance of the models is that they provide a qualitative understanding of the phenomena. A vast amount of research is needed before the models are made general and realistic enough to yield quantitative results which can be relied upon.

8.1. INVERSIONS

The word inversion usually refers to temperature inversions, layers in which the temperature is increasing with height. Usually the temperature decreases upward through the atmosphere to a height of about 10 km. However, inversion layers of limited vertical extent do exist. In such stable layers the mixing is poor since the turbulence is suppressed. Inversion layers aloft tend to function as lids preventing the spreading of effluents to greater heights. There are many reasons why inversion layers may form: Radiation inversions may form over open land, when the sky is clear and the wind weak. At sunset the ground will cool rapidly by the emission of long-wave heat radiation, and the air will be cooled from below; an inversion may also be due to advection of warm air over cold surfaces. Still another type of inversions is caused by the gradual sinking of upper air strata followed by adiabatic heating due to the increase in static pressure experienced by the subsiding air. Such subsidence inversions often occur in connection with stagnant high-pressure systems.

It may be useful to note that from an air pollution point of view it is the variation with height of the potential temperature, which is of immediate importance. It would therefore be appropriate to define an inversion as a state in which the potential temperature increases with height (hydrostatically stable atmosphere).

8.2. TURNING OF WIND WITH HEIGHT – WIND SHEAR

Owing to the Coriolis effect the wind direction changes with height. In the free atmosphere above the boundary layer the pressure forces are roughly balanced by

r the ground, the pre-
wind slows down and
' from the center of a
ld the wind aloft at a
ice roughness and at-
as well as the vertical
since the shears can
1968; Pasquill, 1969;

vays, both by limiting
ersion or by causing
mall standing rotors
lley winds, lee waves
. The meteorological
nge one to one hun-
asingly clear that we
ns of the transport
;o-scale phenomena.

th respect to other
problems.

in along the rim of

(2) Stacks located too close to buildings of approximately the same height as the stack causing down wash in the wake behind the building (See e.g. Slade, 1968; Smith, 1968).

9. Conclusion

The atmosphere is indeed very complex. It swirls around in two and thredimensional 'eddies'. It feeds the chaotic turbulence and is full of organized motion: Convective cells, waves and rolls. All these features are constantly evolving, propagating, break-ing, and dissipating in nonlinear confusion.

Important portions of the meteorological knowledge and experience now available can be applied directly to air pollution problems, especially by those with common sense and observant eyes. A multitude of local air pollution problems could have been avoided had there been awareness of the problems before they became problems.

In general, however, the application of meteorology to air pollution is not always straightforward. It is a task for specialists. With the techniques which are at hand we feel confident that much could and should be done, but we also know that without full and confident co-operation of engineers, architects and city planners meteoro-logical knowledge, however complete, will have little impact on the problems of air pollution.

References

Briggs, G. A.: 1969, *Plume Rise*, USAEC Critical Review Series, TID-25075, 81 pp.
Carpenter, S. B., Montgomery, T. L., Leavitt, J. M., Colbaugh, W. G., and Thomas, F. W.: 1971, *J. Air Poll. Control Ass.* **21**, 491–495.
Changnon, S. A.: 1968, *Bull. Amer. Meteorol. Soc.* **49**, 4–11.
Fiedler, F.: 1969, *Beitr. Phys. Atm.* **42**, 143–173 and 251–296.
Fiedler, F. and Panofsky, H. A.: 1971, *Bull. Amer. Meteorol. Soc.* **51**, 1114–1119.
Forsdyke, A. G.: 1970, *Meteorological Factors in Air Pollution*, Tech. Note No. 114, WMO-No. 274, TP. 153, 32 pp.
Hewett, T. A., Fay, J. A., and Hoult, D. P.: 1971, *Atmos. Environment* **5**, 767–789.
Hino, M.: 1968, *Atmos. Environment* **2**, 149–165.
Johnson, W. B. and Uthe, E. E.: 1971, *Atmos. Environment* **5**, 703–724.
Leavitt, J. M., Carpenter, S. B., Blackwell, J. P., and Montgomery, T. L.: 1971, *J. Air Poll. Control Ass.* **21**, 400–405.
McCormick, R. A.: 1970, *Meteorological Aspects of Air Pollution in Urban and Industrial Districts*, Tech. Note No. 106, WMO-No. 251, TP. 139, 69 pp.
McGuire, T. and Noll, K. E.: 1971, *Atmos. Environment* **5**, 291–298.
Pack, D. H.: 1971, *Meteorology of Air Pollution. Man's Impact on Environment*, McGraw-Hill Book Co., pp. 98–111.
Panofsky, H. A.: 1969, *Amer. Sci.* **57**, 269–285.
Pasquill, F.: 1962, *Atmospheric Diffusion*, Van Nostrand, 297 pp.
Pasquill, F.: 1969, *Phil. Trans. Roy. Soc. London* **A265**, 173–181.
Pasquill, F.: 1971, *Quart. J. Roy. Meteorol. Soc.* **97**, 369–395.
Saltzman, B. E.: 1960, *J. Air Poll. Control Ass.* **20**, 660–665.
Scorer, R.: 1968, *Air Pollution*, Pergamon Press, 151 pp.
Slade, D. H. (ed.): 1968, *Meteorology and Atomic Energy*, USAEC, Div. of Tech. Info., 445 pp.
Smith, M. (ed.): 1968, *ASME Guide for the Prediction of the Dispersion of Airborne Effluents*, 85 pp.
Stern, A. C. (ed.): 1968, *Air Pollution*, vols. I–III, Academic Press, N.Y.

CORROSION PROBLEMS AS A RESULT OF AIR POLLUTION

M. H. TIKKANEN and S. YLÄSAARI

Techn. University of Helsinki, Otaniemi, Finland

Abstract. Air pollution especially by sulphur compounds gives severe corrosion problems. These are discussed and possible means to diminish them. Economic viewpoints are given.

The most important source of increased corrosion due to atmospheric pollution is the existence of ever increasing amounts of sulphur oxides in the atmosphere. This problem is a very old one, but its importance has been realized more widely only during the two last decades. Earlier the attention regarding this problem was limited to localized corrosion phenomena occurring at certain parts, such as economizers and chimneys, at boiler power plants. Later, after the Second World War corrosion damages of this type were multiplied many times over, but at the same time the area of corrosion damages spred outside the boiler plants into their environments, all this due to an overwhelming increase in the use of heavy, sulphur-bearing heating oils.

It is only natural that the significance of the sulphur problem has been recognized in all developed countries. The main attention has, however, been concentrated on its ecological consequenses such as its physiological effect on human beings or its destructive influence on vegetation, and only very slightly on economical losses caused by increased corrosion.

The aim of this survey is to present a modern view about how and why this type of corrosion occurs and works and, further, something about its economical importance and, finally, some thoughts about practical possibilities to prevent it.

1. Basic Facts About Sulphur Dioxide Corrosion

It is an established fact that sulphur dioxide (SO_2) alone does not increase atmospheric corrosion noticeably. Its dangerousness lies in that under favourable conditions it is easily converted into sulphur trioxide (SO_3) which, in turn, forms sulphuric acid with water or its vapor. The sulphuric acid thus formed is the real cause of sulphur dioxide corrosion.

There are two mechanisms by which SO_2 is converted into SO_3 during or after burning of coal or oil:

(1) by reaction of SO_2 with atomic oxygen in the flame,

(2) by secondary catalytic oxidation of SO_2 with molecular oxygen on hot surfaces.

Earlier it was assumed that the oxidation of SO_2 occurred according to reaction (2) only, but, more recently, it has been shown by Johnson and Littler (1963), that the main part of the SO_3 in the flame originates from the reaction with atomic oxygen according to reaction (1). This very important discovery was completely

G. Lindner and K. Nyberg (eds.), Environmental Engineering, 95–101. All Rights Reserved

unexpected since thermodynamic calculations show that SO_3 should be instable at flame temperatures. The explanation for this is twofold, first, that the equilibrium between SO_2 and *atomic* oxygen is different and for SO_3 more favourable than the equilibrium between SO_2 and *molecular* oxygen and, secondly that the dissociation of SO_3 is about eight times slower than its formation. Thus, if the cooling rate of the flame gases is high enough, more or less of the SO_3 formed in the reaction with atomic oxygen will prevail even in the cooled gases in spite of a shift in equilibrium. One factor of greatest importance when determining the stability of SO_3 in flame gases is the amount and quality of solid particles (carbon, ash) in the gas phase. Experience has shown that SO_3 adsorbs extremely strongly on certain fine particles thus increasing its chemical stability considerably.

The abovementioned phenomena can be visualized with the following reactions:

$$SO_2(g) + O(g) \xrightarrow{K_1} SO_3(g) \xrightarrow{K_2} SO_2(g) + \tfrac{1}{2}O_2(g), \tag{1}$$

where $O(g)$ represents gaseous atomic oxygen and O_2 gaseous molecular oxygen. It is evident that amounts of oxygen above the stoichiometric composition increase the percentage of SO_3 in the flame. Thus, one of the most efficient measures in boiler power plants when trying to hold down the SO_3 content is the use of air and oil in stoichiometric proportions.

It has been verified that the SO_3 in the flame gases combines with water vapor into

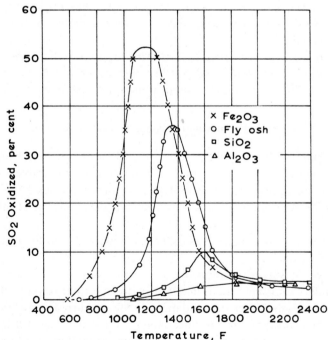

Fig. 1.   Catalytic oxidation of SO_2 to SO_3 by various materials
(according to ASME Research Committee).

sulfuric acid already at or straight below ca. 250 °C. Depending on different factors such as the partial pressure of water vapor-, the temperature and quality of solid surfaces in the furnace and so on, this sulphuric acid will condensate on metallic surfaces at temperatures between 100–190 °C. The sulphuric acid solution thus formed will react with metallic iron into ferrous sulphate a part of which, in turn, will be oxidized into ferric sulphate as follows:

$$2 \ FeSO_4 + SO_2 + O_2 \rightarrow Fe_2(SO_4)_3. \tag{4}$$

Since ferric sulphate is instable at higher temperatures it will be reduced by SO_2 back to ferrous sulfate if the relative contents of O_2 and SO_2 are changed:

$$Fe_2(SO_4)_3 + SO_2 + H_2O \rightarrow 2 \ FeSO_4 + 2 \ H_2SO_4, \tag{5}$$

which then can be oxidized over again. This is the other way to oxidize SO_2 into sulphuric acid continuously with the help of ferrous and ferric ions in solution.

Other forms of catalytic oxidation without the presence of water are also known. Thus ferric oxide (Figure 1) and vanadium pentoxide both act as effectful catalysts capable of oxidizing SO_2 to SO_3. The mechanism of the oxidation is as a matter of principle the same as above, namely the alternating oxidation and reduction of the catalytic metal ions in the solid oxide lattice.

2. Sulphuric Acid Corrosion in Boiler Plants

Practical experience shows that the parts in boiler plants which corrode easiest are the heat exchangers for feed water and for the burning air and the chimneys. In all cases the main cause of corrosion is the condensation of sulphuric acid on solid surfaces. In order to examine this problem in connection with boiler plants it is suitable to treat the subject for two different types of boilers:

(1) water boilers,
(2) steam boilers.

As known, water boilers are mainly used for central heating, air condition and district heating purposes. Since the amount of these boilers is increasing quite fast in Scandinavia and since there seems to be many faults in their construction and use when considering corrosion problems this case will be treated more thoroughly in the following.

It is a common observation that the burners in water boilers are not as effective as in the bigger steam boilers. The result is that the mixing of the fuel and air is incomplete which means a greater tendency to form carbon deposits. This, in turn, leads into situations where unnecessary amounts of air is used thus favouring SO_3 formation. Perhaps the most disadvantageous thing, however, is the lower temperatures of all metallic surfaces due to corresponding lower water temperatures, a fact which in a high degree favours the corrosive sulfuric acid condensation.

Our quite large experience in Finland in examining corrosion damages in water boilers has shown clearly that the worst corrosion cases occur in connection with heavy soot formation on the metallic tubes. These soot deposits are porous and allow therefore condensation of sulphuric acid molecules between the deposits and the metallic surfaces below them. The porous deposits, in turn, form effective insu-lators against the heat transfer from the flames to the metallic surfaces thus pre-venting evaporation of the condensated acid back in the flames. The result is in many cases that boiler tubes with a nominal wall thickness of 5 mm can loose 80% of their material in one year or less, all that due to dissolution in condensed sulphuric acid. It has to be understood that this type of corrosion cannot be prevented by using other tube materials like stainless steel since no ordinary metal is resistant against sulphuric acid at high temperatures. The only practical solution is to change the working procedure in such a way that the soot problem will be eliminated. How this is possible, depends on the local construction and working methods. There are same problems in all steam boilers too but their control is much easier. Usually their automation allows working with stoichiometric air contents by which method the formation of SO_3 can be held at minimum. Since the burners are much more effective and the furnaces much bigger, the formation of soot is no problem as in the earlier case. Corrosion of the economizers is prevented by regulating their tem-peratures just above the condensation temperature thus preventing the damaging condensation of sulphuric acid.

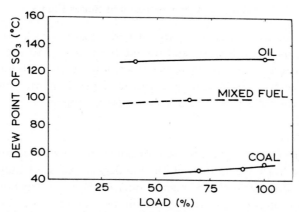

Fig. 2.   The effect of fuel and boiler load on the dew point of the flue
gas (according to Sadik, 1963).

There is an interesting difference between using coal or fuel oil. In the first case it is very easy to prevent condensation of the acid since the ash particles from the burning coal carry most of the SO_3 out of the furnace thus preventing any noticeable condensation of the acid. This has been visualized in experimental trials in big scale in Finland many years ago. When using fuel oil only the condensation temperature was 120–140 °C which dropped to 90–100 °C when 50% of the fuel oil was made up by coal powder or peat, as illustrated in Figure 2.

3. Sulphur Dioxide and Atmospheric Corrosion

As already mentioned sulphur dioxide alone does not increase corrosion but the sulphur trioxide formed by oxidation of SO_2. The role of SO_2 as an increasing factor of the atmospheric corrosion has been explained by Vernon in his classical studies during the thirties. According to him two factors are of primary significance: relative humidity of the air and the content of solid particles, especially that of soot, in the atmosphere. In Figure 3 some of his results are shown.

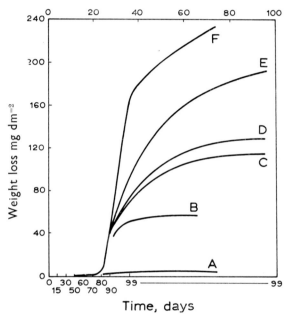

Fig. 3.   A Laboratory study of the atmospheric corrosion of polished steel surfaces
(according to Vernon, 1935).
No SO₂ added: A – activated carbon, B – ammonium sulphate.
0.01 % SO₂ added: C – no particles, D – quartz powder, E – ammonium sulphate, F – activated carbon.

The corrosion damages caused by sulphur dioxide in the atmosphere are numerous and can be divided as follows:

(1) Increased corrosion of unprotected iron, steel and other metals when being exposed to sulphur dioxide containing atmospheres.

(2) Diminished corrosion resistance of different coatings (paints and metallic coatings).

(3) Corrosion damages of ceramic materials.

According to the studies made by the Royal Academy of Engineering Sciences (Sweden) unpainted steel corroded in marine atmosphere about two times and in urban atmosphere about 4 times faster than in rural atmosphere. Due to the catalytic formation of ferrous sulphate during the corrosion of steel even painted surfaces corrode much faster in sulphur dioxide containing atmospheres. Thus in such cases it is necessary to use thicker coatings as specified for more normal cases.

The same results can be shown for other metals and metallic coatings. E.g. the corrosion of nickel and chromium coatings increase in sulphur dioxide containing atmospheres. Zinc coatings are especially sensitive against this type of corrosion. The usual lifetime of hot galvanized coatings in rural atmosphere is about 70 yrs but in urban or industrial atmosphere 7–15 yrs only. Aluminium is extremely resistant against corrosion in rural atmosphere but in severe conditions its use is almost impossible. Thus in industrial locations where the air contains soot or dust together with sulphur dioxide its life-time can be as short as 3–6 months. Even stainless steels are not at all immune against the attack of sulphur oxides. Since these steels protect themselves against corrosion by passivation, which, in turn, implies oxidizing conditions, they are very sensitive against all kinds of deposits.

One of the main areas where sulphur dioxide corrosion has a great damaging effect is that of air conditioning in all of its forms. Increased sulphur dioxide contents in the air mean correspondingly higher contents of sulphuric acid in the cooling waters in the cooling towers or humidifiers. Very often it is possible to find pH-values as low as 3–4 in these waters. What it means in the metallic circulation system consisting parts of steel, aluminium, copper and tin solder, is easily understood.

4. Economic Consequences of Atmospheric Corrosion

For us in Scandinavia the following corrosion costs due to atmospheric impurities, especially those of sulphur dioxide, have been calculated by Bergsman and Liljenvall, Swedish Corrosion Institute Stockholm. As basis for their calculation they used the cost value of Mattsson according to whom the yearly costs for corrosion prevention painting, galvanizing and nickel plating in Sweden was in 1961 about 600 million Swedish crowns. Bergsman and Liljenvall showed that during 1961–1968 these costs increased yearly by about 500 million Swedish crowns. If corrosion damages of non-metallic were included yearly costs increased to 1000 million Swedish crowns.

5. Possible Preventive Measure Against Sulphur Dioxide Pollution

There are three possibilities to diminish the air pollution by sulphur oxides:
  (1) Using low-sulphur oil.
  (2) Cutting down the sulphur content of the crude oils at refineries.
  (3) Desulphurization of flue gases.
There are some types of crude oils with sufficient low sulphur contents but for the time being this possibility is accessible only for a small part of oil users. The second possibility is perhaps the best one with a long-range aim and one has to hope that the big oil companies will do their best in this case. The third possibility has interested many chemical companies and several new patented processes have been published.

So far known only a few of them are in actual use. Their use will always be limited to big plants where the recovery of sulphur has an economic basis.

Summarizing the above mentioned it is clear that the atmospheric sulphur problem regarding its corrosive effect is already of great economic significance even here in Scandinavia. Since its importance has not yet been accepted widely enough the yearly corrosion costs have increased heavily. It is evident that something must be done and soon.

References

ASME Research Committee on Corrosion and Deposits from Combustion Gases Pergamon Press and The American Society of Mechanical Engineers, New York, 1959.

Johnson, H. R. and Littler, D. J. (eds.): 1963, *The Mechanism of Corrosion by Fuel Impurities*, Butterworths, London.

Sadik, Fehim: 1963, *Tutkimus SO₂-pitoisten seoskaasujen aiheuttamasta korroosiosta höyryvoimalaitoksessa*, Diploma work, Institute of Technology, Finland.

Vernon, W. H. J.: 1935, *Trans. Faraday Soc.* **31**.

PART III

AIR POLLUTION

Industrial Examples and Engineering

AIR POLLUTION IN A SULPHURIC ACID PLANT

M. A. SALEH, M. S. FAYED, and M. A. EL-RIFAI

Cairo University, Cairo, Egypt

Abstract. The effects of continuous and intermittent sources of air pollution in a sulphuric acid plant are experimentally investigated. The deviation between time-average concentrations and the prediction of various stack dispersion formulae depends on the down-wind distance and the range of wind velocity. Correction coefficients are suggested for adopting such equations for the estimation of long term pollution.

The solvent of heavy chemical industries generally tending to concentrate at centers in the vicinity of populous areas have pushed some industrially-developing countries to pay serious consideration to water pollution problems. Legislations regulating the emission of obnoxious air pollutants are, however, still lacking in many such countries. Sulphur oxides emitted from the sulphuric acid plant surveyed in this work extended their nuisance to people working inside or living nearby the factory; as well as to animals, vegetation and metallic structures or other installations in the area. The present paper covers the results of the experimental investigation of air pollution around the plant in question, an assessment of the range of applicability of some stack dispersion formulae in describing time-average pollution and an examination of the possibility of their extension for describing long-term pollution.

1. Site-Features

The plant is bounded to the North and North-East by a cactus-infested desert which is in turn limited by 70 000 square meters of trees. The south eastern flank is defined by a rural canal the opposite side of which has vegetation. The western side overlooks a lagoon into which the factory wastes are run and then a desert in which some wireless broadcasting towers stand.

Under normal operating conditions the sulphuric acid absorber tail gas containing SO_2, SO_3, and occasionally some H_2SO_4 mist, is responsible for most of the pollution in the area surrounding the factory. It is continuously delivered to a 60 cm diam. stack discharging 20 m above ground level. Upsets in the converter performance result in large SO_2 concentration transients in the tail gas. Other intermittent sources of emission in the process area include:

– A pyrite-burning furnace where emission takes place both by leakage and by discharging the pyrite ash to the atmosphere.

– An oleum-producing department where SO_3 is emitted when tank cars are charged with oleum; and occasionally through leaks due to insufficient preventive maintenance.

– A ground-level open channel occasionally handling dilute sulphuric acid saturated with SO_2.

G. Lindner and K. Nyberg (eds.), Environmental Engineering, 105–112. All Rights Reserved

2. Experimental

Two series of experiments were carried out in order to investigate the pollution inside and around the factory. In the first series of experiments a number of gas analyses were undertaken at 30 different locations chosen on radial distances from the stack so as to be always in the prevailing downwind directions. The analysis was done with a Kraus portable gas analyser (Kraus, 1955) which is based on the Reich-stop iodine titrimetric method (Fayed, 1968; Jacobs, 1949). The measurements by this apparatus are directly expressible as ppm SO_2. The same instrument has been used in evaluating the effect of the occasional discharge of the weak acid saturated with SO_2.

In the second series of experiments, 23 test sites were selected for installing apparatus of the lead peroxide type recommended by the Department of Scientific and Industrial Research, England (DSIR) (Fayed, 1968; and last reference). Such instruments made possible a continuous long-term comparative evaluation of sulphur oxides pollution; their measurements are declared as mg SO_3/day/100 cm^2 of lead peroxide-impregnated cloth. Of the 23 tests sites 8 were chosen inside the factory and 15 outside it. The fifteen external test locations were selected as to be particularly sensitive to stack pollution; they were chosen as far away as possible from trees inside the factory at unobstructed downwind positions. The other eight instruments were located in the process area at points affected by the discussed intermittent sources thus enacting the consideration of their superimposed pattern in the preparation of pollution contour maps. Data on wind speed and direction, atmospheric temperature, stack gas concentration, temperature and emitted volume were continuously recorded throughout the study which lasted over some six months.

The experimental data thus generated were analysed with a view to:

– Estimating the average degree of pollution inside the factory.

– Comparing test results indicating the effect of wind speed and distance on pollution with the predictions of the Bosanquet-Pearson (Bosanquet and Pearson, 1936), Sutton (Sutton, 1947a, b) and Church (Church, 1949) stack dispersion formulae.

– Preparing topographic pollution maps surveying the area under test for various wind speeds and directions and finding a means of applying correction factors to the Bosanquet and Sutton formulae for expressing long-term ground-level concentrations in the relative units of mg SO_3/day/100 cm^2 of lead peroxide-impregnated cloth.

3.   Results and Discussion

Typical SO_2 concentrations at various points inside the factory are presented in Table I. The data have been taken under an average stack emission rate of 0.5 ft^3. SO_2/s, constituting about 1.14% of the tail gas. Statistical treatment of the data revealed that the degree of pollution in the plant atmosphere averaged around 5–6 ppm of SO_2. Ground-level air analysis indicated that the average pollutant concentrations measured in a time less than five minutes is higher than the time average concentration measured in more than 30 min. The ratio of the two values of measurement

TABLE I

Average degree of pollution

Process area		Inside blower rooms		
Concentration	Frequency	Maximum	Average	Minimum
		11	7.3	3.2
0.1–1	1	Around uncovered discharge stream		
1.1–2	2			
2.1–3	4	Maximum	Average	Minimum
3.1–4	2	220	190	150
4.1–5	4			
5.1–6	4	Inside Hiroschoff furnaces building		
6.1–7	4			
7.1–8	2			
8.1–9	1	Maximum	Average	Minimum
9.1–10	2			
14.1–15	1	14	10	7.2
15.1–16	2			
16.1–17	1	Platform of outdoor		
19.1–20	1	equipment structures		
24.1–25	1	Maximum	Average	Minimum
		8.4	6	0.5

varied from 3 to 10 at relatively low wind velocities and from 1.5 to 14 for moderate wind velocities. Since the probability of occurrence of maximum concentration is another criterion of interest in evaluating pollution hazards, the distribution of the individual point concentration was analysed for two test sites located at 200 m and 150 m, respectively downwind from the stack. With moderate wind velocities the probable maximum concentrations obtained were higher than in the case of light wind speeds.

The effect of an intermittent pollution source has been investigated separately. Figure 1 presents the measurements taken during occasional discharges of the sulphur dioxide-saturated weak acid into the open discharge channel. In these experiments the atmosphere was unstable, wind velocities ranged between 6 and 8 knots and air was sampled at breathing level at distances downwind from the acid discharge point which is at ground level. It can be seen that such pollution transients are quite appreciable.

3.1. COMPARISON WITH STACK DISPERSION EQUATIONS

The ranges of applicability of the Bosanquet-Pearson, Sutton, and Church formulae relating the absolute concentration to distance, wind velocity, stability and stack height, have been tested by a number of gas analyses at selected points, chosen at radial distances of 3.75, 5, 7.5 and 10 stack heights downwind from the stack. The individual sampling periods were of unequal duration ranging from 54 to 372 min.

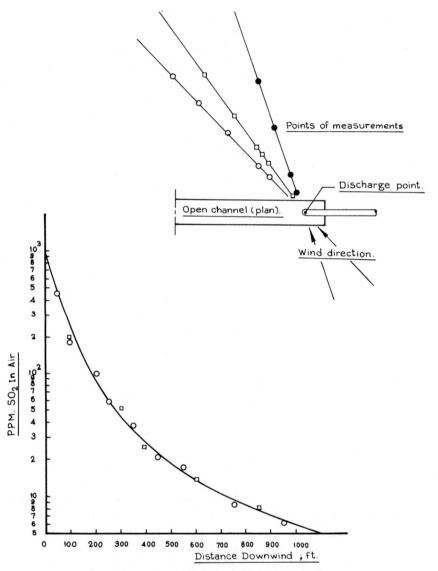

Fig. 1.   Pollution due to weak acid discharge.

with an average sampling period of 235 min. In order to facilitate comparison all the experimental data have been adjusted to correspond to an emission rate of 0.5 ft^3/s of SO$_2$ at atmospheric temperature and pressure. The diffusion parameters and coefficients (Falk, 1952; Magill *et al.*, 1956) used in the application of the above equation were set at values corresponding to the actual conditions of high isotropic instability of the atmosphere. Figure 2 shows how the predictions of the formulae relate to the experimental data obtained for a moderate wind velocity. The lines are

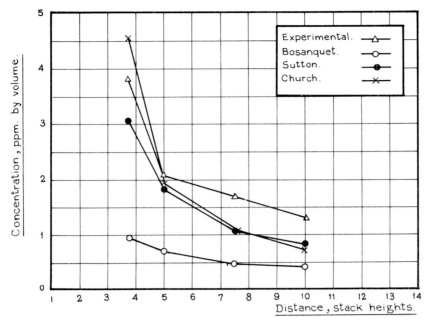

Fig. 2.   Experimental vs calculated pollution at wind speeds in the range of 7–13 knots/h.

shown broken because the average wind velocity for each location is not the same and because the test sites are not all colinear. Table II summarizes the observable trends.

It is seen that the experimental data follow on the average the trend of Church's dilution factors.

3.2. POLLUTION CONTOURS

The results obtained from 23 DSIR probes during several tests periods have been used in preparing pollution contour maps under various wind conditions and stack emission rates. Figure 3 typifies the dependence of the shape of the pollution contour

TABLE II

Agreement between experimental and predicted data

Formula	Wind speed (knots)	Distance stack heights			
		3.75	5	7.5	10
Bosanquet	0– 6	poor	poor	excellent	excellent
Pearson	7–13	poor	poor	poor	poor
Sutton	0– 6	v. good	poor	poor	poor
	7–13	good	good	good	good
Church	0– 6	good	excellent	good	good
	7–13	good	excellent	good	good

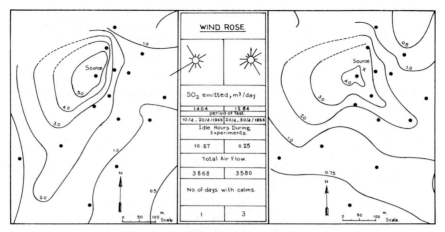

Fig. 3. Typical pollution contour lines.

lines on the wind rose. The analysis of such maps indicated that the rate of SO$_2$ emission from the stack affected the pollution contour mainly outside the process area, except for one site believed to be subject to downwash gases owing to its location beside a relatively high building. Most of the long-term pollution in the process area is otherwise mainly due to intermittent sources.

Long-term stack pollution, measured by lead peroxide probes located between 6 and 35 stack heights downwind has been related to the predictions of the Bosanquet-Pearson and Sutton equations for the corresponding wind speed and weather conditions. For each test period the degree of pollution obtained as mg SO$_3$/day/100 cm^{-2} was divided by the corresponding ppm concentration calculated from each of the two equations. Two corrections coefficients were thus obtained as a function of wind velocity for each location. Such coefficients allow either of the above-mentioned two equations to be applied in estimating long-term pollution contours as measured by the DSIR apparatus. Figures 4 and 5 give the plot of the correction coefficients C_B and C_S to be applied with the Bosanquet and Sutton equations, respectively. It is seen that the correction factors depend more on wind speed than on the velocity down to a minimum in the range of 2–2.5 ft/s, increasing again as the wind velocity increases.

4. Conclusions

(a) Air pollution due to continuous and intermittent sources of obnoxious gases may be investigated by means of relatively simple apparatus. It is possible to obtain fairly representative experimental data on the average potential pollution, the probability of occurrence of peak concentration, and the effects of emission rate, wind speed and direction on the potential pollutant concentration at various points in the area surveyed. A large number of test sites must, however, be chosen in order to take into account the effects of all possible emission sources and of topographic obstructions influencing the pollutant diffusion pattern. The Stochastic nature of the process necessitates, furthermore, a large number of measurements taken at one and the same sampling point and the subsequent statistical analysis of the results.

(b) For this particular plant, intermittent sources were found important in continuous long-term pollution as well as in generating hazardous pollution transients inside the process area. They have been given priority of consideration in the abatement programme.

(c) For predicting downwind time-average pollution, stack dispersion formulae may be used with higher or lower accuracy depending on the range of wind velocities and distance. The application of such formulae is mainly handicapped by the difficulty of obtaining reliable data on wind velocity and direction. In the present plant meteorological data were procured as an hourly average from a recording station situated

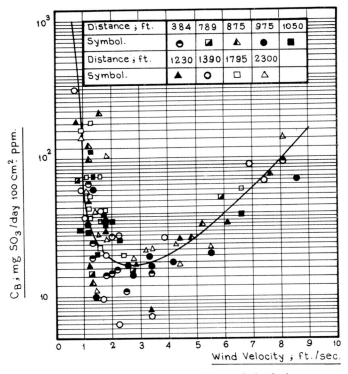

Fig. 4. Correction coefficient, C_B, vs wind velocity.

only a few kilometers from the plant. It was found possible to extend the applicability of such equations for providing an order of magnitude estimation of long-term pollution contours as measured by reagent-impregnated surface probes of the type recommended by the DSIR.

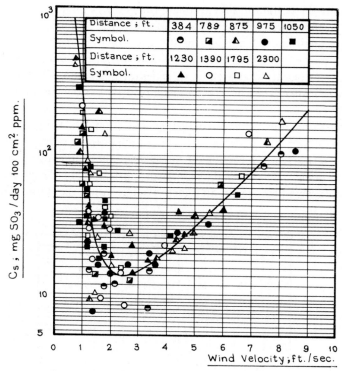

Fig. 5.   Correction coefficient, C_s, vs wind velocity.

References

Bosanquet, C. H. and Pearson, J. L.: 1936, *Trans. Farad. Soc.* **32**, 1249–1264.
Church, P. E.: 1949, *Ind. Eng. Chem.* **41**, 2753–2756.
Falk, L. L.: 1952, personal communication.
Fayed, M. S.: 1968, Thesis, Cairo University.
Jacobs, M. B.: 1949, *The Analytical Chemistry of Industrial Poisons, Hazards, and Solvents* (2nd ed.), Interscience Publishers, Inc., New York.
Kraus, U.: 1955, *Apparatus for the Rapid Determination of Noxious Gases in the Air*, Manual supplied by Strohlein & Co, Ilmenau/Thüringen, GDR.
Magill, P. L., Holden, F. R., and Ackley, C.: 1956, *Air Pollution Handbook*, McGraw-Hill Book Co., New York.
Sutton, O. G.: 1947a, *Quart. J. Roy. Meteorol. Soc.* **73**, 257–281.
Sutton, O. G.: 1947b, *Quart. J. Roy. Meteorol. Soc.* **73**, 426–436.
Manual supplied by Building Research Station of the DSIR, London.

POLLUTION PROBLEMS
IN SWEDISH FERTILIZER INDUSTRIES

BENGT APLANDER

Swedish National Environment Protection Board, Solna, Sweden

Abstract. The bulk of fertilizers in Sweden is produced in Landskrona and Köping. Emissions to water and to atmosphere are accounted for. Different methods to diminish these emissions are briefly discussed with an economical background.

Ammonia and from ammonia produced nitric acid are the raw materials for manufacturing nitrogen fertilizers. The phosphorus in fertilizers derives from rock phosphate, and is made available for the plants through reaction with sulphuric, phosphoric or nitric acid. Potassium in the form of chloride or sulphate can be added in the last part of the manufacturing process in order to get a more complete fertilizer.

1. Emissions

The total emissions from the Swedish works in Landskrona and Köping are presented in Table I. The bulk of fertilizers is produced in these places. The emissions from plants for production of raw materials for fertilizer manufacturing, such as phosphoric acid, ammonia and nitric acid, are included in the given figures, which refer to the state 1969–70.

The emission of plant-nutrients can cause serious water pollution problems at this type of industries. Thus, an increased percentage of phosphate accompanied by higher production of algae has been established. The recipients are, however, influenced by other industrial as well as municipal wastewaters. Emissions of ammonia and its compounds can through a pure poisoning effect lead to death of fish. The

TABLE I

Emissions from two Swedish Fertilizer Industries

Component	Landskrona		Köping	
	to atmosphere kg/h	to water kg/h	to atmosphere kg/h	to water kg/h
Ammonia nitrogen compounds (N)	50	10	45	60
Total nitrogen compounds (N)	100	15	170	115
Phosphorus compounds (P)	–	145	–	2
Fluorine compounds (F)	12	690	0.2	5
Sulphurdioxide (SO_2)	100	–	410	–
Dust	30	–	70	–

G. Lindner and K. Nyberg (eds.), Environmental Engineering, 113–115. All Rights Reserved

emissions to atmosphere consist mainly of sulphurdioxide, ammonia, nitrous gases, fluorine compounds and dust. Especially fluorine can cause damage on certain plants, particulary pine-trees.

2. Methods

A lot of pollution prohibitive measures has already been taken within the fertilizer industry. Thus emissions of ammonia are prevented through absorption in an acid, which often is of clearly economical value. Fluorine in gaseous state is through absorption transferred into a water pollution. Dust is sometimes removed with the help of cyclones and filters, but can in many cases be transferred to a water pollution in a scrubber. In these cases there is often a combined treatment of residual gases containing e.g. fluorine, nitrous gases and dust in a common water scrubber, because the primary task has been to remove fluorine. The reason for this method has been to simplify equipment and thereby lower the investment costs. That gives, however, an unnecessarily high water pollution and this inconvenience can be avoided in different ways, depending on how the gases are emitted. If the dust and the different gaseous pollutants originate from the same source, a scrubber with circulation water is to be preferred. Such a scrubber for 100000 m^3 STP/h is calculated to cost about $\$300000$. The value of the recovered product is supposed to exceed the operating expenses but will cover only a part of the investment costs at an actual installation in a PK-plant.

It also happens that dust loaded air from drying and cooling drums is treated together with residual gases from a reactor. In this case it is preferable to treat the drying and cooling air separately in a filter while the reactor gases are led to a scrubber. The investment cost for an exchange of scrubbers at a Swedish NPK-plant against filters for 200000 m^3 STP/h of dry gases has been estimated to be between $\$400000$ and 600000. The operating expenses of the filters have been calculated to be about $\$120000$/year. This estimated exchange has not been carried through. In order to solve a difficult pollution problem in a delicate recipient it can, however, be necessary to use such methods.

A Swedish urea plant has lately switched over from direct to indirect condensors after the crystalliser. Through the reduced volume of water, it has been possible to distill the ammonia and return it to the process and in this way the pollution of ammonia to water has become a fraction of previous state. The investment cost is about $\$350000$ and the value of recovered ammonia is about the same as the operating costs.

The switch to indirect condensors should be considered in many other connections. The thereby reduced volume of water can either be reused in the process or in another neighbour plant. As a rule it is economically most advantageous to keep the pollutants within the process instead of building external purification arrangements. If a finishing purification is necessary, one should, also in this case, endeavour to get as small volumes of water as possible. External purification methods in full scale are not yet in operation at any fertilizer plant. Among the methods which are object of international research and development work, ion exchange and precipitation of ammonium and phosphate ions, as magnesiumammoniumphoshate ($MgNH_4PO_4 \cdot \cdot 6H_2O$) seem to be most useful.

To reduce the losses of fluorine to the atmosphere in connection with the manufacturing of NPK-fertilizers, the possibility of using evaporated phosphoric acid is considered. This acid contains less fluorine, and shows the influence of the choice of raw material on the pollution problems.

Reference

This report is originally a summary of the results from the work of a committee with representatives from the Swedish fertilizer industry and the National Swedish Environment Protection Board. The completereport is published in Swedish by the National Environment Protection Board: Statens naturvårdsverk publikation: 1971:6, *Vatten- och luftvårdsproblem vid tillverkning av gödselmedel.*

PURIFICATION OF THE TAIL GASES OF NITRIC ACID PLANTS

JORMA SOHLO

University of Oulu, Oulu, Finland

Abstract. Only two percent or less of the nitrogen oxide pollutants are released by industrial production and the use of nitric acid, but local problems in the vicinity of a plant could arise. Absorption
methods, catalytic reduction and adsorption in solids are possible methods in cutting down the NO_x
concentrations in the tail gases. Catalytic reduction is the only commercial method for achieving low
concentrations at the present, but molecular sieve adsorption is likely to compete in the future.

Nitrogen oxides (NO_x) are among the most important air pollutants. About 98% of
them are released through the combustion of fuel in automobiles and power plants.
Only two percent or less are released by industrial production and the use of nitric
acid. However, local concentrations of NO_x in the ambient air can become high in
the vicinity of such plants. Strict legal regulations concerning the NO_x concentrations
are lacking in most countries at the present. In the U.S.A. the law for the ambient air
quality standards to be enforced by 1975 sets the upper limit of the annual mean
concentration of NO_x at 100 μg per m^3 or 0.05 ppm. Standards of this kind make it
possible to calculate the permissible amounts and concentrations released by industry.

1. Concentrations of NO_x in Tail Gases

Absorption of NO_x in water (nitric acid) is the last step in the production of nitric
acid. For economic reasons the NO_x concentrations of the tail gases are relatively
high, i.e. 1000–5000 ppm; 50–70% of this is NO_2 and the remainder mainly NO. The
main component of the gases is nitrogen. In addition, there is some water vapor and
oxygen. The pressure in the adsorption columns is 1–10 atm.

2. Cutting-Down Methods

In order to cut down NO_x concentrations in the tail gases it is possible to employ
absorption methods, catalytic reduction and adsorption in solids. Absorption by
alkaline solutions is possible, but the process easily becomes complicated and expensive if an attempt is made to use the nitrogen bound. Purification to below 200
ppm is hardly feasible by alkaline absorption. In new plants is it possible to reach
700–1000 ppm of NO_x by increasing the absorber capacities at not too high a cost.

2.1. Catalytical reduction

In the U.S.A. it has been customary to reduce NO_2 catalytically to NO. The gases
released are then colourless, but hardly less harmful. By now, however, reduction to
nitrogen is feasible, and concentrations of 100–300 ppm NO_x at the outlet can be
reached. The heat released can be used to pay for the system. The reductant (fuel)

G. Lindner and K. Nyberg (eds.), Environmental Engineering, 117–119. All Rights Reserved
Copyright © 1973 by D. Reidel Publishing Company, Dordrecht-Holland

used is dependent on the local factors. Several plants in the U.S.A. and Europe are employing the method by now.

2.2. ADSORPTION IN SOLIDS

Adsorption in silica gel and active carbon has been an object of research on a small scale, but the processes have not reached a commercial status. Interest in the use of molecular sieves for adsorption has risen in recent years, and a process for adsorbing NO_x is reported to undergo pilot testing in the U.S.A.

2.3. MOLECULAR SIEVE ADSORPTION

The author has conducted pilot-scale research in co-operation with a fertilizer company (Typpi Oy) during 1968–1969, in order to ascertain the usefulnes of the molecular sieve adsorption process. The characteristics of individual sieves can be considerably different. Thus the results given below are typical of the sieve used (Union Carbide, AW-500).

– Typical s-shaped curves were observed at the outlet of adsorption as a function of time ,but before the rise, concentration of NO_x was well below 10 ppm.

– Adsorption capacity was 15–20 g of NO_x per kg of sieve at a pressure of 1 atm. and a temperature of 25 °C, when the outlet was discontinued at 400 ppm. The figure is dependent on the flow velocity as well as the relative and absolute concentrations of NO and NO_2 at the inlet. Inlet temperatures between 25 °C–50 °C also had a slight effect.

– Increase of temperature in the adiabatic operation was slight, and should not give rise to any great technical difficulties.

– Co-adsorption with water was found to take place, and elimination of water vapor is not needed.

– The size of the cylindrical sieves used (1/8″ and 1/16″) had a marked effect only on the pressure drop of the bed.

– Some oxygen is needed to convert NO to NO_2 before adsorption takes place.

– Desorption with air was already taking place at 150 °C. The concentrated gas mixture can then be recycled to the absorption columns of nitric acid manufacture.

The experiments showed the process to be operable on a small scale, and no factors were found which could prevent full-scale operation. The practicability of the process is thus dependent merely on the economic factors, which of course are connected to the technical ones.

The capital costs of the process are relatively high, since at least two columns are needed for continuous operation. Costs in excess to those of catalytic reduction could be expected.

The operating costs are mainly dependent on the price and the lifetime of the sieve. The capacity of the sieve investigated was rather low, but the same manufacturer reports double capacity for a newer type. This would mean 3–4 w% of NO_x at 1 atm. Nitric acid plants using higher pressure are in a more favorable position, since the adsorption capacity of molecular sieves is a function of pressure. Another manufacturer gives a capacity of 8–10% of NO_x at 8 atm. pressure and a lifetime in excess of 1000 cycles of sorbing and desorbing. The prices of the sieves vary around 3 \$ per kg. Improvements to these figures are to be expected.

A very preliminary cost estimation on the basis of these figures shows that increased production of nitric acid could probably pay for the operating costs at 1 atm. pressure. As plants normally operate at a higher pressure, it seems likely that amortisation would be possible too.

3. Conclusions

Comparison of the prospects for different processes is somewhat uncertain. The usefulness of a process depends on the requirements set for the output concentrations of NO_x in the exit gas. This in turn depends on the standards for ambient air quality, size of the plant and height of the exit pipe, among others. If an exit concentration of 700–1000 ppm is allowed, an additional absorption section in new nitric acid plants could be the most economical answer. Other absorption methods have some promise of slightly lower values.

Catalytic reduction is the only commercial method for achieving low concentrations at the present. In the near future, molecular sieve adsorption is likely to appear as a competing method. An ambient air concentration of 0.01 ppm of NO_x caused by a factory, would require exit values under 50 ppm for the tail gases of nitric acid plants. Under such conditions, a molecular sieve adsorption system might prove the most economical approach to purification.

References

Newman, D. J.: 1971, *Chem. Eng. Progr.* **67**, 79.
Sundaresan, B. B. and Harding, C. I.: 1968, *Methods of Recovering Gases and Vapors*, U.S. Patent 3,389,961.
Sundaresan, B. B., Harding, C. I., May, P. P., and Hendrickson, E. R.: 1971, *Environ. Sci. Technol.* **1**, 151.

DUST-ABATEMENT IN THE CEMENT INDUSTRY

MARTIN FUREVIK

F. L. Smidth & Co A/S, Environmental Technology Division, Copenhagen, Denmark

Abstract. In the cement industry the dust control equipment is integrated in the manufacturing process, therefore it is especially important in this industry. Dust problems arising during cement manufacturing and handling and storing of materials are accounted for as well as the four main types of filters – cyclones, scrubbers, fabric filters, and electrostatic precipitators.

Without suitable dust control equipment, large dust emissions will occur at several stages in the conversion of limestone and clay into Portland cement. The dust problems are so large that 5–10% of the total cost of a new cement plant will have to be spent on measures for the prevention of air pollution.

However, in the cement industry the dust control equipment is integrated in the manufacturing process. Without dust filters 20–30% of the production would be lost, and one can imagine what that would mean to the production costs. But admittedly, keeping up with most emission regulations makes it necessary with more efficient (and more expensive) filters than pure economic considerations would lead to.

It is here the intention, first to describe the four main types of filters and then show how they are applied in the cement industry.

1. Dust Collectors

1.1. CYCLONES

Cyclones (Figure 1) are the simplest and cheapest type of dust collector commonly used. Their operation is normally troublefree, and the main operating cost is the power required to overcome the pressure drop of the cyclone. This is usually in the range 50–150 mm WG.

The dust-laden air enters through a tangential inlet, and a powerful vortex is created in the cyclone. The strong centrifugal force pulls the dust out of the vortex and towards the wall. The dust then moves to the conical outlet while the clean air leaves by the central outlet.

Large cyclones are efficient in removing dust larger than 10–30 μm. A higher dust collection efficiency may be obtained by using a multicyclone, which is a parallel arrangement of several small cyclones (0.15–0.30 m$\varnothing$). The centrifugal force is greater in multicyclones than in large cyclones, because the radius is smaller, and the efficiency is therefore higher; most of the particles larger than 3–5 μm are collected.

Most multicyclones are, however, more prone to wear than large cyclones, and they also have a tendency to plug when the dust concentration is high.

1.2. SCRUBBERS

If a cyclone cannot give a sufficiently high dust removal efficiency, a scrubber may be

G. Lindner and K. Nyberg (eds.), Environmental Engineering, 121–127. All Rights Reserved
Copyright © 1973 by D. Reidel Publishing Company, Dordrecht-Holland

Gas outlet

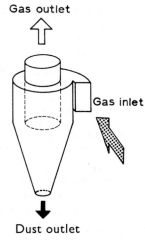

Gas inlet

Dust outlet

Fig. 1.   Cyclone.

the best solution in some cases. A scrubber is a dust collector where the dust is caught by water drops which are later on removed from the air stream. Many types exist, generally they have a low first cost, but high operating costs due to a high power consumption; the most efficient ones usually have the highest power requirements.

1.3. FABRIC FILTERS

Fabric filters are usually made in the form of large vertical bags of fabric suspended from the roof of the filter casing as shown in Figure 2. When the air passes through the fabric, most of the dust is deposited on the fabric surface. The bags are cleaned at regular intervals by reversing the air stream or shaking the bags. The dust then falls down in the hopper in the form of large agglomerates.

A breakage of one filter bag will cause a large increase in the dust emission. How-

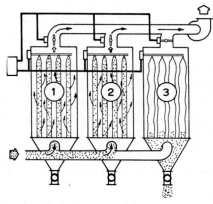

Fig. 2.   Fabric filter. The filter bags in section 1 and 2 are in operation while the bags in section 3 are cleaned.

ever, the advent of filter materials made out of synthetic fibers has made breakages far less frequent than earlier when natural fibers were used. A well designed filter is made up of several sections. If one of the filter bags in one section breaks, that section can be shut down for repair, while the remaining sections take the whole load. However, this will to some extent increase the risk of breakages in the other sections.

Fabric filters are limited to 300 °C operating temperature, and the filter materials which can be used above 200 °C are expensive.

1.4. ELECTROSTATIC PRECIPITATORS

Properly applied the electrostatic precipitator gives highly efficient gas cleaning at lower operating costs than any other type of dust filter. Large capacity precipitators may also have somewhat lower first costs than bag filters, while smaller precipitators generally are more expensive than corresponding bag filters. However, an electrostatic precipitator only functions effectively if the resistivity of the dust (i.e. the specific resistance of the dust) lies between certain limits.

The dust laden gas passes between two sets of electrodes (Figure 3), the discharge

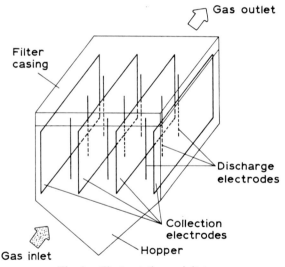

Fig. 3.   Electrostatic precipitator.

electrodes which are made out of thin wires and the collecting electrodes which are made out of steel plates. The voltage between the electrodes is so high (40–80 kV) that the discharge electrode will emit electrons which will wander towards the collection electrode. The dust particles pick up charge from this electron current and will therefore also move towards the collection electrode where they form a dust layer. The collection electrodes are periodically rapped. This causes large agglomerates of dust to fall down in the hopper below the precipitator.

2. Dust Problems Arising During Cement Manufacturing

2.1. GENERAL

As the wet process and the dry process are by far the most common ways of making cement, the dust problems of other methods will not be dealt with.

The raw materials are first crushed and ground to a very fine size before being burnt to clinkers. The clinkers – nodules about 25 mm in diameter – are then cooled, mixed with 5% gypsum and ground to cement.

In the wet process a slurry of limestone and clay is ground, while in the dry process, if there is any water present, it will be evaporated before or during grinding. The processes are similar after the kiln.

The dust problems arising at each step of the processes will be dealt with in turn. Afterwards the dust problems associated with handling and storing of materials will be described.

2.2. CRUSHING

Only when the raw materials have a low water content does crushing give rise to dust problems.

The dust released during crushing is prevented from escaping by fitting a hood to the crusher and drawing air in through the hood. It is quite common, but seldom satisfactory to use cyclones for filtering this air. Exit concentrations from cyclones are usually 1–2 g Nm^{-3}, while less than 0.1 g Nm^{-3} can be obtained by use of fabric filters.

For the rare combination of dry raw materials and the wet process, a scrubber may be used with advantage, as the water discharged from the scrubber can be used later in the process.

A precipitator cannot be used for this dust as it has too high resistivity.

2.3. RAW MILL

The air stream from a dry process raw mill contains large quantities of dust; 60–90 g m^{-3} from mills with mechanical raw meal transport and 500–750 g m^{-3} from air-swept mills. The air temperature is 100°C, and the drying of the raw meals has made the air moist.

At this stage some remarks have to be made about how the resistivity of the dust is influenced by the gas temperature and humidity.

Figure 4 shows that the maximum resistivity occurs in the range 180–250°C. Increase of humidity means decrease in resistivity at temperatures less than 350°C, while above 350°C the influence of humidity is negligible. With a dew point of 10°C, the resistivity is high even at temperatures below 100°C. Thus electrostatic precipitators cannot be used for filtrating cement dust entrained in ambient air. Figure 5 shows that the resistivity also depends on the type of dust. The humidity has a similar influence on these curves as on those in Figure 4.

Thus it is seen that the dust from the raw mill is ideal for a precipitator. The raw

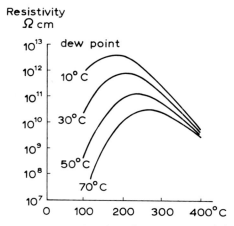

Fig. 4. Resistivity as a function of temperature and dew point.
(Here shown for the dust from a long dry kiln.)

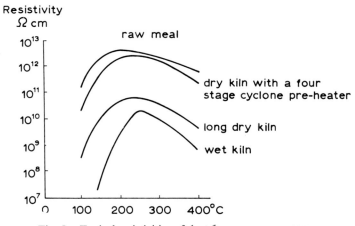

Fig. 5. Typical resistivities of dust from some processes.

mills are usually so large today that fabric filters become a more expensive alternative.

The hot air used for drying the raw meal in the mill is taken from the kiln exit or an auxiliary furnace. If the kiln exit gas has been used in the raw mill, the gas will usually be passed through cyclones before being returned to the kiln filter. In this way the cost of one precipitator is saved in addition to not having to use fuel in an auxiliary furnace. Most of the dust is caught in the cyclone, while the filter at the kiln catches almost all the remaining dust.

2.4. ROTARY KILN

This is the largest dust source in a cement factory. The gas volume is large, and the dust concentration is high (Table I). The filter at the kiln usually costs more than all the other filters together.

Up to this point only the dry process has been dealt with. In a wet process kiln, however, the slurry is dried, and dust problems at once arise

The dust from a wet kiln is, however, easy to deal with in an electrostatic precipitator. The dew point is so high that the resistivity of the dust is always sufficiently low. Insulation of the filter casing may, however, be necessary in order to avoid condensation and corrosion.

TABLE I

Emissions from cement kilns

Type of kiln	Temperature at exit from kiln	Gas flow Nm³/kg clinker	Dust concentration at kiln exit g Nm⁻³	Percent of dust less than 3μm
Wet kiln	120–350	2.8–4.5	10–50	6–23
Dry kiln	350–450	1.4–3.0	20–100	45–85

It is not possible to return all the collected dust to the kiln as this will cause an increased concentration of alkalies (mainly K_2O) which will form deposits in the kiln or deteriorate the cement quality.

By using a precipitator it is possible to enrich the alkalies 4–5 times in a portion of the dust, and the remaining dust – about 90% – can be returned to the kiln. This is because the alkalies are collected mainly in the second half of the precipitator from where it can be taken out by a separate screw conveyor.

The alkali rich dust may be used as a fertilizer, but is usually disposed of in an empty pit.

Although fabric filters and even scrubbers have been used at wet process kilns, these do not have the capability of enriching the alkalies in one portion of the dust.

As can be seen from Figure 5, the dust from dry process kilns has a high resistivity. This is especially the case with the most common type at new plants, namely the kiln with a four stage cyclone preheater for raw meal. The most satisfactory way of collecting the dust from such a kiln is first to cool the gas to 150 °C (by evaporation of water in a large tower) and then to pass it through an electrostatic precipitator. Treated in this way, the dust will have a suitable resistivity for collection in a precipitator (Figures 4 and 5). If an evaporation tower is not used, the precipitator will have to be 2–4 times larger which means a higher total cost.

An alternative is to pass the gas through the raw mill before passing it to the precipitator. The drying of the raw meal gives the gas a suitable temperature and humidity for filtration in a precipitator. However, as the kiln is in continuous operation while the mill may be stopped one day per week, an evaporation tower will be necessary part of the time.

Fabric filters are usually more expensive for this application than precipitators as they will have to filter large volumes of hot air. (Usual values are 10^5–10^6 m^3/h at 300 °C).

As dry kilns operate with little excess air, explosive concentrations of carbon monoxide may occur. A CO-measuring instrument which can switch off the precipitator current is therefore always fitted.

2.5. CLINKER COOLER

The airflow through a clinker cooler of the grate type is too large to be used entirely for combustion in the kiln. Part of the air must be filtrated and discharged to the atmosphere. An outlet concentration of 0.15–0.25 g Nm^{-3} can be achieved by using high efficiency cyclones. These must be wear resistant as the dust is very abrasive. A fabric filter or a Kiesbett filter (where the air is filtered through a layer of gravel) is used if lower emissions are required.

However, the planetary cooler represents the ideal solution to this pollution problem as all the air passing through the cooler is used for combustion. In this way one saves a filter in addition to obtaining better heat economy.

2.6. CEMENT MILL

Large amounts of heat are released during grinding in cement mills, and water cooling by atomizing nozzles is commonly employed. The air temperature at the mill outlet is usually 100 °C, and the dew point is 60 °C. This makes the dust ideal for a precipitator while fabric filters may be prone to clogging.

3. Dust Problems Arising During Handling and Storing of Materials

Materials are transported by conveyor belts, bucket elevators, and pneumatic transport. Eliminating the dust which these create may not have any significant effect upon the total emissions of the plant, but may mean a lot to the indoor working conditions.

Close fitting hoods are used at the discharge and receiving ends of bucket elevators and conveyor belts while the rest of the devices may be completely enclosed. Fabric filters are almost always used for filtering the air which is drawn off from the hoods. The dust has too high resistivity at ambient temperatures for precipitators and is usually too fine for cyclones.

Fabric filters are also used for venting pneumatic transport devices.

If a stockpile with dry raw materials or the clinker store is not enclosed, large amounts of low level dust may occur in windy weather. A silo or a closed storehouse is recommended. Fabric filters may be used for venting these buildings.

Acknowledgement

The author acknowledges with gratitude the assistance of J. H. Lind, Chr. Petersen and E. Ødum in the preparation of this paper.

AIR AND WATER POLLUTION PROBLEMS
AT PETROLEUM REFINERIES

BO ASSARSSON

Swedish National Environment Protection Board, Stockholm, Sweden

Abstract. The petroleum refining capacity in Sweden will probably increase from some 13 million tons of crude oil in 1970 to an annual capacity of about 40 million tons at the end of the seventies. A committee formed with members from the Swedish Petroleum-Refining Industry and the Environmental Protection Board has investigated the situation of pollutants and different abatement methods.

The National Swedish Environment Protection Board has initiated investigations of the pollution problems of different industry branches. This work has been and is carried out in co-operation with the industry. In the same manner a committee has been formed with members from the Swedish Petroleum-Refining Industry and the Environment Protection Board in order to investigate the pollution problems of the refineries.

1. Situation at Present

There are five petroleum refineries in Sweden with a total capacity around 13 million tons of crude oil per year. The two biggest plants, the BP Refinery and the Shell-Koppartrans Refinery, have annual capacities of 5 million tons each and are both situated in Gothenburg. They produce primarily petroleum gases, gasoline, kerosine and fuel oils.

Nynäs-Petroleum has one refinery in Nynäshamn, south of Stockholm, producing around 3 million tons per year of primarily gasoline, kerosine, fuel oils, bitumen, and lubrication oils. Nynäs-Petroleum has also bitumen plants in Gothenburg and Malmö.

2. Futural Increase

Scanraff, an affiliated company to OK, consumers co-operative oil-company, is building a refinery of 7 million tons annual capacity at Lysekil on the Swedish west coast. This will be in operation in 1974. The Shell-Koppartrans Refinery at Gothenburg has been permitted by the Franchise Board for Environment Protection to extend to maximum 13 tons per year capacity. BP has at the Franchise Board applied for permission to extent their refinery at Gothenburg to maximum 15 tons per year capacity. Finally, Nynäs-Petroleum has announced plans for an extention or a new refinery. Summarized, these plans would lead to an annual refining capacity at the end of the seventies of around 40 million tons approximately covering the expected consumption of oil products at the time. This situation also brings good possibilities for government action in favour of fuels with low sulphur and lead-content.

3. Petrochemical Industry

The Swedish petrochemical industry is situated at Stenungsund on the west coast. Esso Chemical has there a steam cracker producing around 250000 tons of ethylen per year. The ethylene production in Sweden at the end of the century has been prognosticated to between 2.5 and 4 million tons per year. This would mean an extention of the industry at Stenungsund and one or two new steam-crackers at existing refineries on the west coast.

4. Pollutants

4.1. AIR POLLUTANTS

The potential sources of air contaminants at petroleums refineries are summarized in Table I.

The total emission of sulphur oxides from the Swedish petroleum industry was in 1970 10000 tons. The sulphur content in the fuel used at the different plants varied from 0.4 to 2.0% calculated as mean values including fuel gas. Burning of off-gases from a sour-water stripper gives an emission of 1200 tons of sulphur dioxide at one

TABLE I

The potential sources of air contaminants at petroleum refineries

Type of emission	Potential source
Sulphur oxides	Boilers, tail gases from sulphur recovery, catalyst regenerators, decoking operations, flares, heaters, incinerators, acid sludge disposal
Hydrocarbons and odors	Air blowing, barometric condensers, blowdown systems, boilers, catalyst regeneration, compressors, cooling towers, decoking operations, flares, heaters, incinerators, loading facilities, processing vessels, pumps, sampling operations, tanks, vacuum jets, waste-effluent-handling equipment
Carbon monoxide	Catalyst regenerators, flares, coking operations, incinerators
Nitrogen oxides	Boilers, catalyst regenerators, flares, incinerators
Particulate matter	Boilers, catalyst regenerators, coking operations, heaters, flares

plant. Burning of tail-gases from sulphur recovery plants emit 1400 tons of sulphur dioxide to the atmosphere. These figures should be compared with the total sulphur dioxide emission in Sweden that has been estimated to 870000 tons in 1970.

The emission of hydrocarbons and odors is very difficult to calculate, but a rough estimate gives a figure around 7000 tons per year.

The emission of nitrogen oxides has been calculated from emission factors given in literature. Measurements confirming this has also been carried out. The emission factors used are 2.5 kg nitrogen oxides per ton of fuel gas and 8.5 kg nitrogen oxides per ton of fuel oil. This gives an estimated emission of 2300 tons of nitrogen oxides per year from the Swedish petroleum refineries.

The emission of particulate matter is very difficult to calculate or measure but it is very obvious that this problem is of minor importance at this industry.

4.2. WATER POLLUTANTS

The different types of aqueous effluents that arise at petroleum refineries can be divided into the following groups:
  – water from processes,
  – polluted cooling water,
  – unpolluted cooling water,
  – polluted rain water,
  – unpolluted rain water,
  – water from cavern storages,
  – drainage water from tanks,
  – ballast water,
  – sanitary waste water.

It is clear that these aqueous effluents contain a wide range of organic and inorganic pollutants. Information on the quantities of pollutants is known for the Swedish refineries but data concerning the effluents from the different unit processes are very scarce. A few data on this is given in literature. Waste water from crude desalter amounts to 3–9% on crude. The concentration of sulfides range from 0–13 mg/l, phenols from 10–24 mg/l, oil from 20–516 mg/l and BOD 68–610 mg/l.

Other effluents from processes are mainly sour condensates originating from condensed steam that has been used to facilitate distillation and cracking operations. According to API data the concentration of pollutants in these sour waters are the following:
  – H_2S ranging from 300–11 000 mg/l,
  – NH_3 ranging from 100–7000 mg/l,
  – phenols ranging from 100–1000 mg/l.

Data from the Swedish refineries show great variations in effluent quantities.

The older refineries use large volumes of cooling water, around 2000 m^3/h, while modern plants discharge almost no cooling water. The quantity of waste water from processes are more equal but also here newer refineries are more favorable.

The most common equipment used for treatment of effluent water at the Swedish refineries is at present gravity-oil separators. A sour-water stripper is also used at one plant and a biological treatment facility at another, both treating process water. An estimate of the total emission of water pollutants from the refineries would give the figure 50 kg oil/h, 2.5 kg phenol/h and 200 kg BOD/h.

5. Abatement Methods

5.1. AIR POLLUTANTS

Possibilities to further reduction of the pollution problems of petroleum-refineries exist. The emission of sulphur dioxide from burning of fuels can be reduced by decreasing the sulphur content of the burned fuel oil, increasing the fuel-gas burning or by introducing desulphurization of combustion gases. The sulphur content of the fuel oil can be reduced by burning the heavy fraction of a low sulphur crude or by burning desulphurized fuel oil.

Since desulphurization of heavy fuel oil has not been applied yet at Swedish refineries the sulphur content of the fuel oil can only be reduced by increasing the portion of low sulphur crude. Because of the higher cost of these crudes and the different yields of e.g. gasoline, costs up to 400 U.S. $ per ton of sulphur not emitted have been reported. This is a strong incentive to develop and install desulphurization processes for heavy fuel oil for which costs around 180–240 U.S. $ per ton of sulphur captured have been reported. Gasification of heavy oil is another possible way to reduce the sulphur dioxide emission at costs around 240 U.S. $ per ton of sulphur. Finally, development work on combustion gas desulphurization processes is carried out in many places around the world today. It is expected that several processes will be ready for general application on a large scale in the seventies and at costs competitive to fuel oil desulphurization.

The emission of sulphur oxides from burning of tail gases from sulphur recovery plants represents an increasing problem when the capacity of desulphurization processes at refineries increase.

At the usual type of sulphur recovery, the Claus-kiln, the efficiency in recovery is about 95%. This efficiency can be increased to 97% by installing a third step in the Claus-kiln. The sulphur oxides can also be removed by using a gas desulphurization process. Because of the high concentration of sulphur oxides this can be done at rather low cost by using available technique.

The emission of hydrocarbons and odors can be reduced by different technical methods, e.g. floating roofs on tanks, gas recovery systems, mechanical seals, vapor incineration, but a very important factor is a proper maintenance of compressors, pumps, valves, and other equipment.

The technical measures to prevent emission of particulate matter are the same as at other combustion equipment. One exception is the combustion of waste gases in flares, where steam injection is used to prevent soot formation.

5.2. WATER POLLUTANTS

The emission of water pollutants can be reduced by several principally different methods. Improved efficiency of processing is one way to reduce the problem. Extraction agents are available which are more selective and can often be regenerated. Probably the most important improvement has been the application of hydrodesulphurization for removal of sulphur components from distillates. The previous methods used alkali, caustic soda, solutizers, sulphuric acid, or oleum which created water pollution problems.

Improved drainage systems with separation of oil-free rain water and other types of aqueous effluents facilitate purification processes and prevent pollution of oil-free drainage water.

Cooling water can be reduced by using cooling towers with circulating cooling water or by using air cooling for most of the products. The possible reduction is from

3000 m³ water per hour for 1 million ton per year crude oil throughout to ca 20 m³ per hour.

Process water can be reduced, e.g., by using steam condensates as water for desalting crude oil and for using steam for special purposes. Refinery effluents that have been purified to a high degree can be used, if the salt content can be kept sufficiently low, for make up water in circulating cooling system of for preparing boiler-feed water.

Stripping of sour waters, i.e. process waters containing sulphides and other volatile material is carried out with flue gases or with fuel gas and steam in a tower. The off-gases are burnt or led to a sulphur recovery unit.

The next step is usually the application of gravity-oil separators in which oil accumulates on the surface and solids with some absorbed oil settle to the bottom. The commonly used separator is designed by the American Petroleum Institute (API). Under unfavourable conditions the oil content in the effluent of the API separator may leave something to be desired. An improvement has been obtained with the parallel plate interceptor and the newer corrugated plate interceptor.

If further removal of oil and suspended solids is required, chemical flocculation or air flotation and different biological purification methods can be applied. The choice and combinations of these processes is dependent on local circumstances and costs. The effect depends on many factors, e.g. the quality of the processed water, but requirements for oil content less than 2–3 mg/l and phenol content less than 0.2 mg/l have been met.

CALCULATION OF MIST ELIMINATORS
FOR INDUSTRIAL GASES

ULRICH REGEHR

EUROFORM, Aachen, Germany

Abstract. Mist elimination from flowing gases is an important field, although this is not fully recognized yet. Mist eliminators are used for process gas cleaning and exhaust gas cleaning. Different types of eliminators, their functions as well as their possible installations, are accounted for. The article ends with an example of calculation.

Mist elimination from flowing gases is a relatively new specialised field of which little is known in industry. For this reason it is generally confusing for the buyer, when planning installations and applications, to make the right choice amoung the different types of mist eliminators. Since mist elimination is a field of growing importance, the engineer with his technical background can realize an optimum in the planning of installations and applications. With a better understanding he can utilize the data given by the producer of mist eliminators.

1. Application of Mist Eliminators

Mist eliminators are required for the elimination of liquid drops or hanks from flowing gases. This is done for two reasons:

1.1. PROCESS GAS CLEANING

The gas is removed from the liquid as it would effect the further use of the gas.
  Examples:
  – Elimination of liquid hydrocarbons during the benefication of natural gas.
  – Elimination of sulfuric acid in the production of chlorine gas.
  – Elimination of water from air in air conditioning installations.
  – Elimination of condensates of hydrocarbons from crude gas in polyolefin installations.

1.2. EXHAUST GAS CLEANING

The gas is removed from the liquid since liquid contributes to environmental pollution.
  Examples:
  – Elimination of scrubbing fluid containing fluor from exhaust air produced by the aluminium melting process.
  – Elimination of acids entrained on aspiration of galvano-technical baths.
  – Elimination of scrubbing fluid after the scrubbing of exhaust gas in fertilizer plants (scrubbing of nitrous gases).
  – Elimination of water from exhaust air in cooling towers.
  – Elimination of scrubbing fluid in the phthalic acid scrubbing process.

TABLE I

Data for the calculation of a mist eliminator

No.	Data	Knowledge is necessary for	Example
1.00	Variable of state of gas		
1.10	Gas flow	Dimensioning of mist eliminator	210.000m³ h⁻¹
1.20	Composition	Selection of material, corrosion	Exhaust air from phthalic acid plants
1.30	Temperature	Selection of material	40°C
1.40	Pressure	Case-construction and liquid-drain	1 atm. approx.
1.50	Density	Calculation of mist eliminator	1.2 kg m⁻³
1.60	Viscosity	Calculation of mist eliminator	approx. 15.6 × 10⁻⁶ m² s⁻¹
1.70	Humidity	Deciding if the drops can evaporate (built-up)	Saturated
2.00	Variability of state of the liquid		
2.10	Loading	Dimensioning of mist eliminator	25 m³ h⁻¹
2.20	Composition	Selection of material, valuation of built-up danger	Water, enriched with 1–18% of organic acid
2.30	Density	Calculation of mist eliminator	approx. 10³ kg m⁻³
2.40	Drop size distribution	Calculation of mist eliminator	Expected distribution: < 65 μm
2.50	Pollution, e.g. solid particle	Evaluation of danger of built-up and clogging, corrosion	When falling short of saturation, phthalic anhydride is produced
2.60	Surface tension	Evaluation of wetting behaviour	like water
3.00	Operation data		
3.10	Admissable pressure drop	Dimensioning of mist eliminator	max. 100 mm water gauge
3.20	Admissable installation area	Dimensioning of mist eliminator	3.500 mm = height 3.200 mm = width
3.30	Desired efficiency	Dimensioning of mist eliminator	99.9 % for drops > 14 μm 60 % for drops = 8 μm 40 % for drops = 6 μm
3.40	Velocity distribution in front of mist eliminator	Dimensioning of mist eliminator	horizontal flow after pre-eliminator (length = 4 m)
3.50	Limit drop size	Dimensioning of mist eliminator	as small as possible
3.60	Direction of flow	Selection of mist eliminator type	horizontal

To make complete and optimum calculations for mist eliminators with specific requirements the supplier must have a range of values like those summarized in Table I.

Frequently, only incomplete specifications are outlined by the buyer of a mist eliminator. This applies especially in specifying the drop-size distribution. In such cases,
the supplier is dependent on his own experience and judgement.

2. Synoptic Table of Mist Eliminator Types

Mist eliminators presently on the market can be summarized in the following groups:
Table II and Figure 1.

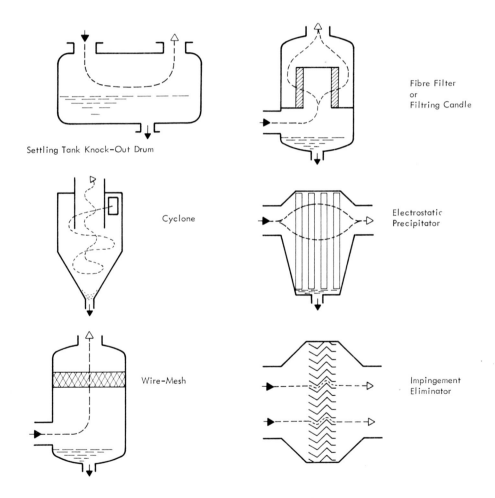

Fig. 1.   Mist eliminator skeletonized.

TABLE II

Synoptic table of Mist Eliminator types

Name	Operating mechanism	Assembly	Function	Remarks
Settling Tank Knock-Out Drum.	Gravity	Tank (upright or lying) without installation.	Slowing down of the gas flow so much that the sinking velocity is higher than the gas velocity.	Simple, inexpensive apparatus for very gross drops, poor efficiency, rare use.
Fibre Filter Filtering candle.	Filter mechanism	Case with compact packing of fibres or with different single or individual filtering candles.	Single drops are brought in touch with the fibres, flow together, increase and fall down due to gravity. The filtering candle prevents passing of droplets while gas flows through.	Voluminous apparatus with very low face velocity and low liquid loading preferably for very small drops, danger of clogging and built-up.
Electrostatic Precipitator.	Electrostatic powers	Case with electrodes.	The drops are electrocharged and eliminated along the collecting electrodes.	Complicated, expensive apparatus for extremely fine drops, very high efficiency, rare use.
Cyclone	Mass moment of inertia	Tank with installation which force the gas flow to rotate.	Based on rotation separation because of varying density, the drops are eliminated along the tank walls.	Simple and voluminous apparatus for middle-sized drops, good efficiency, frequent use.
Wire Mesh	Mass moment of inertia	Tank with pack of several layers of wire mesh of undulated wires, compact wire mesh with high porosity.	The drops are brought in touch with the wire surface, flow together, increase and fall down due to gravity.	Voluminous apparatus with low face velocities for low liquid loading and very fine drops, danger of clogging and built-up.
Impingement eliminator	Mass moment of inertia	Tank with a set of profile plates of different types.	Gas flow is split up into many single flows and repeatedly deflected, due to inertia the drops cannot follow the flow of the gas and thus are eliminated on the impingement surfaces.	Small type of construction because of high face velocity, very high efficiency even for very fine drops, low pressure drop on well formed profiles, increasing use.

3. Detailed Viewing of the Impingement or Flow Grid Eliminator Types

Mist eliminators based on the effect of inertia are most important for industrial mist elimination. Cyclone and wire mesh eliminators have long been widely used in industry. Less so, has the impingement eliminator. The present type model of this eliminator is relatively new and, because of its advantages, it is increasingly gaining importance. Therefore, the following analysis is confined to the impingement eliminator.

4. Explanation of its Function

With the impingement eliminator the gas flow is divided into single flows and repeatedly deflected; see Figure 2.

Within the deflection, inertia forces act on the drops producing relative motion between gas-flow and the path of motion of the drops. If the inertia forces are large enough, the drops reach the impingement surface, thus completing the primary process of mist elimination. The liquid, resting on the impingement surfaces, then has to be led out of the eliminator. This process-termed secondary elimination is performed by the phase separating chamber as seen in Figure 2. This type of eliminator is especially developed for horizontal gas flow. The liquid is pushed towards the phase separating chambers of the eliminator by the inertia forces of the gas.

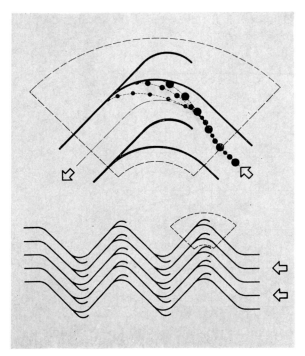

Fig. 2. Diagrammatic representation of mist elimination in
an impingement eliminator for horizontal flow.

Within the range of the phase separating chambers the forces become inactive and the liquid, under the influence of gravity, runs into a collecting tank. In eliminators with vertical flow the liquid is collected, also under the influence of gravity, on the inlet edges of the eliminator. From there it runs down in big drops or hanks.

5. Possible Installations of the Impingement Eliminator

Three possibilities exist for the spatial arrangement of the impingement eliminator.
  (1) *Horizontal installation:* Vertical gas flow.
  (2) *Vertical installation:* Horizontal gas flow.
  (3) *Sloping installation with angle to the vertical:* Gas flow with the same angle to the horizontal.

Due to the influence of gravity and the drainage of the liquid, the construction of a mist eliminator operated as in point (1) has to be different from the construction of those eliminators operated with a gas flow as in point (2).

6. Calculation of the Elimination Process

The laws of impingement eliminators can physically be determined. By means of a mathematical representation of the flow field in the eliminator, the path of motion of a single drop can be calculated. From this, explicitly results the so called limit drop size d_T^*. This, by definition, is the diameter of the smallest drop which, after coming

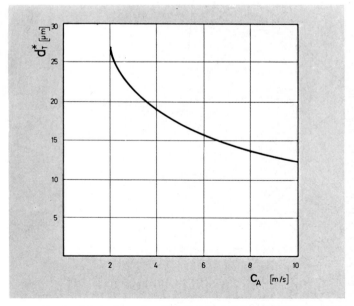

$$d^*_T = k \left(\frac{\varrho_G}{\varrho_T} v_G \frac{1}{c'_A} \frac{r_a - r_i}{\varphi} \right)^n$$

k, n : constants
ϱ_G : density of gas
ϱ_T : density of liquid
v_G : viscosity of gas
r_a }
r_i } : measurements of flow grid
φ }
c'_A : gas velocity in the flow grid
c_A : face velocity

Fig. 3.   Limit drop size, d^*_T, for the EUROFORM high efficiency mist eliminator TS 5 (System water-air, 20 °C, 1 ata).

from the most unfavourable position in the eliminator and flowing through the deflection, reaches one of the impingement surfaces. The equation in Figure 3 comprises
all values which are critical for the mist elimination in an impingement eliminator.
Term 1 defines the variable state of gas and liquid, term 2 the dimensions of the
eliminator, and term 3 the gas velocity and, thus, the gas flow. With the given law the
efficiency of an eliminator can now be calculated provided the size of drops to be
eliminated, hence the drop size distribution, is known. If the calculated limit drop size
is smaller than the smallest occuring in the drop size distribution, the efficiency theoretically is 100%.

Figure 3 shows the mathematically determined limiting drop size d_T^* as a function
of the face velocity c_A for a particular mist eliminator. With a face velocity of 7 m s^{-1},
for example, the drop limit size is 14 μm. These data are relevant to water drops in
an air current at 20 °C and a pressure of 1 ata. Consequently, if a mist eliminator of
this type is operated with a face velocity of 7 m s^{-1} its efficiency theoretically reaches
100% provided all drops in the air current are larger than 14 μm. If, in a drop size
distribution, there also exists smaller drops than the drop limit size d_T^* the elimination
efficiency can be calculated by means of the fractional efficiency, ε^*, Figure 4. The
curves shown in the diagram are for two different types of mist eliminators and for a
face velocity of $c_A = 5$ m s^{-1}. From this diagram the following can be deduced:

 – Drops with a diameter of 20 μm will be completely eliminated.
 – Drops with a diameter of 8 μm will only be eliminated to appr. 60% in the type
TS 5.

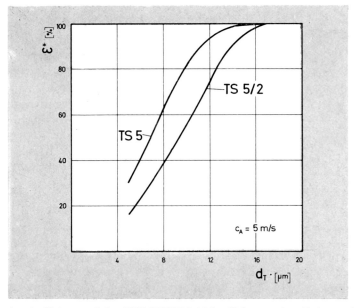

Fig. 4.   Fractional efficiency of the EUROFORM high efficiency mist eliminator TS 5, TS 5/2
(System water-air, 20 °C, 1 ata).

7. Expenditure of Energy in Elimination

The expenditure of energy in elimination is represented by the pressure drop Δp.

$$\Delta p \approx \zeta c_A^2$$

ζ: resistance coefficient of the mist eliminator, c_A: Face velocity.

The resistance coefficient is a characteristic of apparatus and can be held relatively low with a favourable design of the impingement surfaces profiles. From the relation it follows that an increase of the face velocity c_A results in a squarely increasing pressure drop, Δp. The face velocity should therefore be chosen as low as the given drop size distribution will allow in order to keep the operating costs which are connected with the pressure drop as low as possible.

8. Calculation of a Mist Eliminator for a Given Exhaust Gas Problem

The following demonstrates the course of calculation necessary to determine the eliminator in a given exhaust gas problem. Hereby it is assumed that the supplier disposes of a type-program from which the suitable eliminator for the actual application can be chosen. For this course of calculation it is referred to the example given in Table I

Point 3.60 in Table I states that the flow direction of the exhaust gas is horizontal. Consequently, a type of eliminator is chosen that is suitable for horizontal flow. The desired efficiency is given in point 3.30. As it is required that all drops with a diameter larger than 14 μm are to be eliminated by 99.9%, the drop limit size d_T^* may not exceed 14 μm. The face velocity necessary for this drop limit size can be taken from the diagram in Figure 3. The result is a drop limit size of $d_T^* = 14$ μm and a face velocity of $c_A = 7$ m s^{-1}.

Point 3.30 requires an efficiency of 60% for drops of 8 μm and 40% for drops of 6 μm. Figure 4 shows that these values can be obtained with the chosen eliminator type TS 5. The diagram is valid for a face velocity of 5 m s^{-1}; yet, since the chosen face velocity is 7 m s^{-1} the required values for the efficiency are surpassed.

The dimensioning of the eliminator can be ascertained by the volume flow of 210000 m^3 h^{-1} corresponding to 58.2 m^3 s^{-1}, as in point 1.10. From the relation

$$F_A = \frac{Q}{c_A} = \frac{58.2 \text{ m}^3 \text{ s}^{-1}}{70 \text{ m s}^{-1}},$$

follows a face area

$$F_A = 8.3 \text{ m}^2.$$

The max. safe installation area is 3.5 m $\times$ 3.2 m = 11.2 m^2. Consequently there is enough space for the installation of an eliminator with frame construction.

The admissable pressure drop of 100 mm water gauge, point 3.10, is also within the allowable range. Figure 5 shows for the mist eliminator TS 5 a pressure drop of appr. 65 mm water gauge with a face velocity of 7 m s^{-1}. Thus, a reserve of 35 mm water

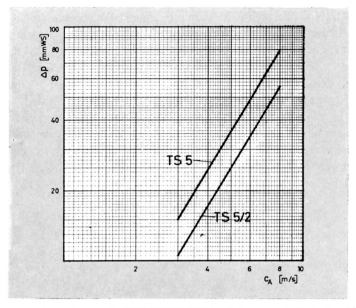

Fig. 5.   Pressure drop, Δp, for the EUROFORM high efficiency mist eliminator TS 5, TS 5/2
(System water-air, 20°C, 1 ata).

gauge remains for interference effect on the pressure drop due to case and frame construction.

The material is chosen according to the relevant tables for chemical resistance and the composition of the gas (1.20), the temperature (1.30) and the composition of the liquid (2.20). The selected material is stainless steel (No. 1.4571). The gas temperature of 40°C, point 1.30 would permit the use of a plastic eliminator which is essentially less expensive than one of stainless steel. Examinations of chemical resistance however, prove that not enough knowledge exists in the use of plastic. Furthermore, point 2.50 indicates a possible build-up of the eliminator. Cleaning would then become necessary and cause heavy mechanical stress which the plastic might not stand.

For the selection of the face velocity c_A the following should be considered: A max. safe face velocity exists for each eliminator. Exceeding this max. velocity would cause some of the liquid already eliminated to come back, and thus a rapidly decreasing efficiency would be the result. For the selected type eliminator TS 5 the max. safe face velocity under normal conditions lies between 16 and 20 m s^{-1} for the water/air system. The chosen face velocity of $c_A = 7$ m s^{-1} thus provides a safety distance of min. 2.3. Such a safety range, even if not always in the same order, is highly important because of the velocity distribution of the gas flow, as stated in point 3.40. In most of the cases the velocity distribution is irregular and velocity peaks occur. If a velocity peak exceeds the max. safe face velocity for the mist eliminator, liquid will flow out of the eliminator, and thus the efficiency will rapidly decrease. This is one of the reasons, frequently overlooked for the failure of mist eliminators.

If the eliminator is operated in a vacuum or under high pressure, the drop size limit has to be calculated according to the equation in Figure 3 with modified density (1.50) and modified viscosity of the gas (1.60). As compared to the normal system, the max. safe face velocity has to be calculated in like manner. The face velocity decreases with the increasing density of the gaseous phase, but when the density is decreasing, the admissable face velocity is increasing, such as when vacuum operated.

The buyer of a mist eliminator will, in most of the cases, not govern its calculation. He should, however, request that the supplier states the drop limit size in the efficiency-guarantee, the max. safe face velocity as well as the safety range for the chosen face velocity. Through the knowledge of these values the buyer obtains a higher degree of safety for the performance of the mist eliminator.

PART IV

AIR POLLUTION

Analysis and Monitoring Control

AIR POLLUTION MEASURING NETWORKS
WITH REMOTELY CONTROLLED SO$_2$ MONITORS

S. M. DE VEER

N.V. Philips Gloeilampenfabrieken, Eindhoven, The Netherlands

Abstract. In October 1969 a network system was put into operation in the Rijnmond area near to Rotterdam in order to measure the air pollution – especially SO$_2$. The results have been successful, indeed. A nation-wide network in the Netherlands is now being installed. The system principles and advantages are discussed.

Although the Netherlands is only a small country, there are living many people, whilst a lot of industry is spread over a number of industrial areas.

The high density of population and industry influences the environment to a large extent. Air pollution, for instance, is a large problem in this country. For this reason the Dutch Government decided some time ago to install a large nation-wide network to measure the air pollution throughout the country, in order to be able to take short and long term measures.

This network is now under construction. Besides, networks in Italy, Switzerland, and Sweden have been installed or are under construction.

A regional network has been put into operation in October 1969 in the Rijnmond area, near to Rotterdam. A large part of this paper will describe this Rijnmond system, because already two years of experience have been accumulated with it.

1. What Measures can be Taken to Prevent Air Pollution?

Obviously the best measures are those that prevent the emission of pollutants. Examples of such measures are the changes now being introduced in the design of car engines, waste gas burners in industrial installations, etc. However, these measures are not capable of eliminating all air pollution. The complementary solution is to prevent the emission of pollutants during periods of stable weather conditions, and to defer polluting activities to such time that the stable weather conditions have ceased to exist. This approach reduces air pollution levels without paralysing economic life. Generally these producers require a limited amount of engineering changes and a great deal of organisation. For example, waste products may be stored during periods of atmospheric stability, and discharged when the wind and vertical convection have returned. Other possible measures are the use of fuels with low sulphur content and postponing cleaning actions during periods of atmospheric stability.

In order to establish such an air pollution management system, however, complex procedures have to be set up by means of extensive regional, national or even international coordination, standardisation of maximum acceptable pollution levels, adequate legislation and effective monitoring.

G. Lindner and K. Nyberg (eds.), Environmental Engineering, 147–157. All Rights Reserved
Copyright © 1973 by D. Reidel Publishing Company, Dordrecht-Holland

Fig. 2a.   Philips type PW 9700 SO₂ monitor in weather/childproof cabinet with door open.

Figure 4 shows the central control room of the system, with the computer proper on the right hand side and the topographic display against the rear wall.

The nucleus of the Rijnmond system hardware is a Philips P 9201 computer with a 12 K,16 bit core memory and a real-time clock. Via a multiplexer the computer is connected to the data receiving side of the MTT system.

Fig. 2b.   Philips type PW 9700 SO2 monitor in weather/childproof cabinet with inside covers removed.

General purpose peripherals connected to the computor include a logging type-writer, a dialogue typewriter to print calibration values and errors, a highspeed paper tape punch, a high-speed paper tape reader, and a topographic display of detector sites to allow real-time surveillance of the region. For special purpose monitoring a flat bed analog recorder is available.

5. Rijnmond Software

The 31 SO_2 measurements signals and 2 fixed test signals are scanned directly.

Every hour, mean values of SO_2 concentration are computed for every site in the area. The hourly mean values are typed-out for immediate investigation and punched-out to allow off-line processing for trend studies, etc.

Fig. 3. SO₂ monitor cabinet with the 'sniffer'.

Fig. 4. The central control room of the system.

Wind direction data also arrive at the central control room in the form of pulse signals. These, however, are converted for analog realtime recording and indication. The wind direction indicator generates digital sine and cosine signals, which are stored in the computer and integrated over one-hour intervals. The resulting values are used to label the hourly SO_2 readings.

The mean values of SO_2 concentration will be different from each other and not intercomparable. This is due to two factors. Firstly, a shift of mean wind direction alters the relative position of the sampling station with respect to the main sources of pollution. Secondly, the diurnal and seasonal variations in stability and the living habits of the population – heating of homes – may cause a systematic variation of mean concentration. For this reason, each hourly mean value is labelled with three indices, one related to site (1 through 331), one related to wind direction (1 through 8 for each of the compass octants, 0 for variable wind or 9 for failure) and one related to time of day (0 through 23). Further, each labelled hourly mean value is taken relative to its own mean – that is, the mean of hourly mean values determined at the same site, at the same wind direction, at the same time of day. This provides 'reduced' values, Δ_i, which are intercomparable:

$$\Delta_i = \frac{C_{ijk}}{\bar{C}_{ijk}} - 1,$$

where C_{ijk} = concentration measured at site i with wind direction j at hour of day k; $\bar{C}_{ijk}$ = a mean value of all values C_{ijk} from the last three months with equal i, j and k.

The great advantage of this concept is that the reduced values can be averaged. Geographical and topographical factors are practically eliminated, so that the reduced area mean value can only be affected by stability in the atmosphere. This is true provided that the emission rates of the various industries remain statistically constant during long periods. Air pollution concentrations are thus converted into a metrological parameter, and accurate predictions become possible.

Consequently, the reference value, $\bar{C}$, may be considered typical for a specific detector, at a specific time and with a specific wind direction. Reduced values, Δ_i, are typed-out for observation. Furthermore, each hour the eight highest values, Δ_i, are selected and their mean value, Δ_8, tested against a preset upper threshhold value.

This value has been calculated by Dr. L. A. Clarenburg of Rijnmond and represents a certain statistical evidence (Clarenburg, 1968a, b). When this is exceeded an 'internal' pollution alarm is given, both visually and by print-out (by night the control room is not attended and the operator is then automatically warned by telephone; see paragraph on control room). Every 24 hrs, daily mean values are printed for each sensor and a punch-out is made of the entire memory contents to enable stored data to be preserved in case of failure.

Error and failure check programs are included to protect keypoints of the main program and the inputs. Abnormally high or low measured values are rejected and excluded from the calculations of mean values, $\bar{C}$. Detected malfunctions are printed in code by the dialogue typewriter.

Every thirteen hours a 30-min SO_2 detector calibration program is started. Thirteen hours are chosen, instead of twelve, to ensure that the resulting interruption of monitoring is always at a different time of day. To perform the calibration program the computer generates command signals which are directed to the MTT system for transmission to all SO_2 detectors during the calibration cycle. Command signals, received by a detector during the first 10 min of the program, set and hold the selection valve of each SO_2 monitor in the 'zero' position. The measured value, now supplied by the monitor, represents the zero-current of the instrument. On receipt by the computer the zero-value is stored for use as a measurement basis during the next measurement period.

Command signals, received by a monitor during the subsequent 10 min of the program, set and hold the selection valve in the 'calibrate' position. Measured data now attain a value which depends on the zero-value and the quantity of SO_2 supplied by the SO_2 source inside the detector. After subtraction of the zero-value a figure is calculated, indicating the relationship between measured values and a scale factor expressed in $\mu g \ m^{-3}$ of SO_2 per count value. This scale factor is also stored to calibrate measured data over the next measurement period.

After completion of the two 10-min periods command signals are stopped. Now the selection valve automatically returns to the 'measure' position. During the first 10 min of measurement ($=$ the last 10 min of the calibration program) measured data are not processed; this avoids errors while the detector assumes its normal condition. The automatic return of the valve to the 'measure' position when command signals stop ensures that the detector remains operative in case of a transmission interrupt, e.g. at the end of the calibration program, and that line interferences do not cause any disturbances.

6. Central Control Room

The measured data, received from the SO_2 detectors, are processed on-line to compute a general 'pollution figure' Δ_8. When this figure exceeds a preset threshold, 'internal' pollution alarm is given by the computer. This alarm indicates, basically, that meteorological conditions are favorable for the build-up of atmospheric stabilisation and, consequently, also for an increase of the immission.

With an 'internal' alarm the central control room operator asks for the regional weather forecast. If the information anticipates continuation of the prevailing meteorological conditions for at least six hours, the immission, especially of malodourous compounds, is expected to attain an unacceptable level. Consequently emission must be reduced. To achieve this an 'external' alarm is released to industries in the region. On receipt of the alarm predetermined rectifying action will be taken by all concerned. These actions are on a voluntary basis by industry, as a legal requirement is not yet in force. Results of the approximate forty alarms given during the first two years of operation of the network are positive – both with respect to industrial action and to network performance.

To sound an 'external' alarm, advantageous use is made of the Semaphone system

– developed in the early sixties by the Dutch telephone authorities and Philips; meanwhile extended throughout Benelux.

In the Rijnmond region 'pollution officers' have Semaphone receivers with a common call-number. Consequently the control room operator only needs to dial the corresponding telephone number to alert everybody concerned, wherever he may be. Six codes are used to give an indication of the nature of the alarm.

On a topographic display panel, each SO_2 detector is represented by a lamp. This lamp is illuminated when the mean value of concentrations measured during the last 12 min. exceeds a threshold value preset by hand. This facilitates continuous real-time surveillance. Any detector may be selected for a continuous recording of its measured data on a flat-bed recorder. By means of these records and the display the source of an abnormal emission can be 'sensed'.

This facility may, for example, be used when complaints are received from the public (the Rijnmond control room incorporates a special telephone service for this purpose). Local investigations can also be performed by mobile units, operating independently to the network.

The central control room need not to be manned continuously. The operator on duty can insert his private telephone or Semaphone number into a 'Rehalarm' installation connected to the computer. In case of an alarm this installation repeatedly dials the number until the call is answered.

7. Experiences of the First Years of Operation

During the first two years of operation about forty times an alarm has been given to industry. As mentioned before, these alarms are given when weather conditions are beginning to stabilize and are predicted to persist during longer than 6 hrs. Normally the number of telephone complaints by the local population to Rijnmond Authorities increases enormously during such a period. However, each time after an alarm has been given the number of these complaints decreased significantly. This proves how effective the measures, taken by industry, can be.

Besides these very positive results, serious air pollution sources have often been detected by the system, using a combination of SO_2 concentration and wind direction data to determine a position from crossbearings. The identification of such a source not only gives rise to decrease in air pollution but sometimes also saves money for the industry when inefficient burning or a heavy leakage had not yet been detected by the personnel.

Another very important, not precalculated effect of the system is the favorable change of attitude of chemical industry personnel towards the air pollution problem by the repeated communications with the Rijnmond Authorities.

8. General Philosophy of the Dutch National Network

Contacts between the Dutch National Institute of Public Health and Philips led to two

DRAFTPLAN NATIONAL
AIR-POLLUTION MONITORING SYSTEM
Provincial/Regional condensation

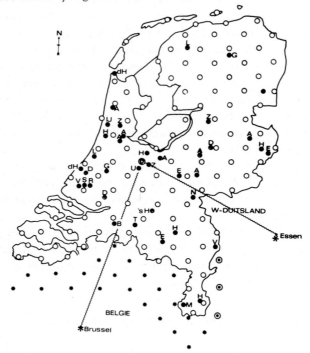

○ Stations in the "grid"
● Extra stations in large cities
◉ National Institute of Public Health in Bilthoven
◉ Linked station of the West German network
∗ Linked station of the Belgian network

Fig. 5.   Draftplan national air pollution monitoring system. Provincial/regional condensation.

developments. Firstly, the detailed design and realisation of an SO_2 monitor along the principles described. Secondly, the development of a nation-wide air pollution monitoring network in the Netherlands. A map (Figure 5) of the Netherlands shows clearly that the idea is to have one national centre, located at Bilthoven near Utrecht, connected to a base line network of monitors with a shortest distance between them of 20 km. This network has a number of local increases in density and totals around 220 monitors.

The function of the network is threefold, viz.:

(1) To measure actual air pollution levels for the identification of critical concentration of pollutants for instant alerting purposes.

(2) To establish the geographical distribution of the air pollutants.

(3) To establish trends in the potential air pollution levels for long term prediction purposes.

The first function is to be performed by regional networks. Each regional network will be controlled from a relatively simple online real-time data processor, which is connected to the monitors. These regional data processors interrogate each monitor once per minute, and establish a mean hourly value out of the data obtained from each measuring site. They signal possible excess of preset limit values on the basis of which action may be taken; for instance, industry may be requested to defer particular activities. The latter functions are entrusted to the national centre. Alerting is only considered a necessary function in a number of highly industrialised areas. Outside these areas only information for the latter functions is required, and there the density of monitors will be much lower.

At the end of 1973 the network will be completed for SO_2 measurements. Then the system will be extended to measure also other components like CO, NO, NO_2, O_3 and H_2S.

References

Clarenburg, L. A.: 1968a, *A Telemetered System to Predict Unfavourable Weather Conditions*, Paper 68–55, presented at the 61st Annual Meeting of the Air Pollution Control Association, Saint Paul.

Clarenburg, L. A.: 1968b, *Precision of Atmospheric Sampling for Air Pollution Levels in Cities and in Industrial Areas*, Paper 68–42, presented at the 61st Annual Meeting of the Air Pollution Control Association, Saint Paul.

De Veer, S. M., Brouwer, H. J., and Zeedijk, H.: 1969, *Remotely Controlled SO₂ Monitor for Operation in Air Pollution Networks*, Paper, 69–6, presented at the 62nd Annual Meeting of the Air Pollution Control Association, New York, June 22–26.

GAS-ANALYTICAL SUPERVISORY CONTROL IN AIR POLLUTION

KURT W. HECKER

Hartmann & Braun AG. Measurement and Regulation Dept., Frankfurt/Main, Germany

Abstract. The measuring tasks are accounted for. They are emission controls of sulphur dioxide and nitric oxides, emission measurements on motor vehicles, immission control by measuring stations and mobile measuring stations as well as ambient air control. Measuring methods like infrared analysis, flame ionization detection, and electro-chemical methods are briefly discussed with practical examples.

1. The Measuring Tasks

1.1. GENERALITIES

For the degree of industrialization of today, including the steadily increasing traffic on the roads, the problem of avoiding air pollution is becoming more and more important. Gaseous and dust pollutions in the atmosphere has reached such an extent that suitable sweeping measures with the aim of limiting the emission of pollution have become absolutely necessary.

Investigations have shown that atmospheric pollution is mainly based on three groups of emittents, namely industry, domestic fuel, and road traffic. Further, the degree of pollution can be influenced by certain meteorological factors, such as wind conditions, thermal structure (inversion layers) and others have to be considered, being liable for the propagation and the local accumulation of the harmful substances. In order to resolve the problem of air pollution, data concerning the degree, origin, propagation, and effect of the atmospheric pollutions are necessary. In order to obtain data, reliable measuring instruments must be used. The advantages of continuous, automatic, and recording measurement must surely be self-evident.

From the technical point of view of measurement there are two different types of tasks: emission and immission measurement. By emission measurement pollution concentrations can be determined at the generator, i.e., the emittor; the effect of pollution emission on far-away soil and air layers is determined by immission measurement. The concentration values to be determined and the analyzers to be used yield the emission measurements, meeting the volume percentage range, while immersion concentrations will meet the ppm range (10^{-4} vol. %) and the ppb range (10^{-7} vol. %) due to the high rarefaction of the impurities in the ambient air.

For a number of pollution components, such as carbon oxides, sulfur oxides, nitric oxides, and hydrocarbons, as well as for a number of pollution emittors such as industry firing and vehicles, public regulations already exist to limit pollution to pre-determined maximum concentrations including their monitoring.

Besides the above-mentioned gas-analytical monitoring of open air, control has to be extended to the measurement of the concentration of ambient air, containing substances injurious to health. It means controlling isolated rooms, where people are

G. Lindner and K. Nyberg (eds.), Environmental Engineering, 159–187. All Rights Reserved
Copyright © 1973 by D. Reidel Publishing Company, Dordrecht-Holland

working, such as work-places, garages or tunnels, being endangered by unhealthy gases.

1.2. EMISSION CONTROL OF SULPHUR OXIDES AND NITRIC OXIDES

In order to avoid air pollution the measurement of the SO_2 concentration of is very important. SO_2 is found as an exhaust component in all kinds of firings using sulphureous combustibles as well as in chemical processes converting products based on sulphur. Small amounts of SO_2 in the air can be injurious to the organism. Eyes can already be irritated by SO_2 concentrations of about 20 ppm, while amounts going up to 50 and 100 ppm easily lead to difficulties in breathing. Under certain conditions, such as high moisture and dust content of the air, these symptoms can already be observed for much lower concentrations. Even lower concentrations are harmful for plants: a SO_2 concentration of 0.02 ppm disrupts the assimilation process and hinders photosynthesis.

As a consequence of increasing industrialization and the demand for energy and the erection of new industrial plants and power plants, the emission of polluting substances such as SO_2 has risen enormously during the last decades. According to an investigation in 1962 the total amount of emitted sulphur oxides for one year was 3.6 million tons for the Federal Republic of Germany. Similar values would probably be true for other industrial countries. Public and industrial power plants, factories and housefirings are mainly responsible.

In order to avoid increasing pollution of the air, as well as injury to people and nature, including damage to plant life, different measures are being put into operation in order to diminish SO_2 concentrations. In regions with favourable wind conditions a sufficient rarefaction of emission concentrations in the air can be obtained by building higher smoke-stacks, thus purefying the lower air layers. Further measures to diminish SO_2 concentrations can be different chemical proceedings of ad- and absorption; proceedings with low temperature of flue gas have the disadvantage that the flue gas is pulled down to the bottom, right next to the stack, due to lack of thermal lift.

For some time now public laws and regulations have been in existence with the object of limiting SO_2 emissions to proscribed maximum values. The verification of the efficacy of these emission limitations necessarily requires suitable measuring instruments, for which binding instructions were elaborated. These instructions also refer to the minimum requirements which are to be set for the qualification test for emission measuring instruments as well as for corresponding institutes, and further publication of details of suitable measuring apparatus, their construction, and their operation.

One of the officially approved measuring installations for coal and fuel firings corresponding to the regulations of the Federal Republic of Germany is the measuring instrument URAS EM-SO_2. This apparatus is complete, from the gas sampling probe up to the analyzer.

The sampling place for the gas analysis, assuring a representative analysis, is principally chosen by the competent board of control. In principle the sampling place should be chosen before the flue gas drain enters the stack, i.e., behind filter units, as the durability of the preparation system for measuring the gases is lengthened. For emittors with more than one firing plant with different SO_2 concentrations and having a common stack, special sampling techniques must be developed. In this context we shall emphasize the possibility of using a bypass to the stack, which leads the flue gas to be analyzed to the measuring installation after having passed the mixing zone.

The schematic assembling of the measuring plant is shown in Figure 1. The gas flow is sucked out by means of a sampling probe with a built-in exchangeable ceramic filter. In order to avoid condensation the probe filter is heated up by an electrical heating collar to a temperature which is higher than the acid dew point of the flue gas. The function of the probe heating can be controlled by means of a current relay. A constant dew point of the measuring gas is adjusted in the following electrical measur-

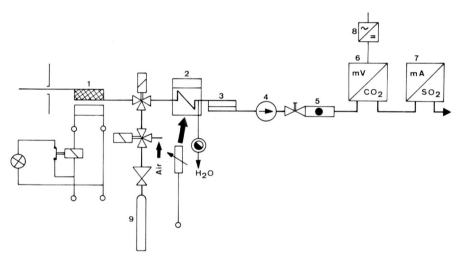

Fig. 1. Measuring arrangement for SO_2 emission. 1 – sampling filter with heating device; 2 – cooler; 3 – membrane filter; 4 – pump; 5 – flowmeter with needle valve; 6 – analyzer for CO_2; 7 – analyzer for SO_2; 8 – power supply unit; 9 – test gas.

ing gas cooler. Here steam influence on the measurement is eliminated and even a following condensation in the diaphragm pump and in the membrane filter is avoided. In cases where there is a high rate of condensation the application of an automatic drain device for condensed water is indicated. The measuring gas cooler can also be equipped with a built-in temperature sensor to control the cooling temperature; this facilitates the localization of measuring errors due to fluctuations in the steam content in the gas sample. The necessary fine filtering of the gas sample in order to protect the measuring systems of the analyzers is obtained by a membrane filter. An electrically operating diaphragm pump forces the measuring gas through the analyzers into the exhaust gas line passing a flow meter. Clean ambient air is used to adjust the zero point of the analyzers. Test gas from bottles is used for end-point control. In principle all test gases are fed in before the measuring gas cooler, because the test gases should have the same dew point as the gas sample.

An infrared analyzer is used for measuring the SO_2 concentration. The minimum measuring range is 0–2 g SO_2 m^{-3}. To realize eventual secondary air inbrake into the flow gas system, diminishing the true SO_2 concentrations in the exhaust gas, a CO_2 thermal conductivity analyzer with a measuring range of 0–20 vol. % or an oxygen analyzer with a measuring range of 0–5 vol. % can be used.

The whole installation except the sampling probe and the recorder located in the

Fig. 2.   Measuring equipment for SO₂-emission.

control room should be built into a rack (Figure 2). This guarantees sufficient protection of the measuring instruments and of the accessories and most of the servicing work at the measuring installations can be effectuated at the same place.

In order to avoid errors by eventual secondary air intakes it is reasonable to refer the limit value of the admissible sulphur content in the fuel to its carbon content by setting up the rate of concentration of SO_2 and CO_2. Conclusions can be drawn from the sulphur content in the fuel on the premises of using the same fuel.

Among the harmful substances besides sulphur dioxide in the atmosphere special attention should be payed to the nitric oxides NO and NO_2. Even though official determinations proscribing an emission control on nitric oxides in combustion installations, do not yet exist in the Federal Republic of Germany, corresponding official regulations can surely be expected in the next few years.

An efficient restriction of the emission of nitric oxides resulting from combustion can only be obtained by extensive constructions and process engineering modifications at the boilers. The existence of people interested in measuring installations for emission of nitric oxides is justified because of these aspects.

Manufacturers of boilers used in power stations and garbage combustion plants as well as manufacturers of metallurgical kilns are already operating nowadays with experimental plants for testing future constructions. On the other hand the legislator charges institutes and laboratories with the elaboration of principles for future official regulations.

For the determination of nitric oxides different measurements can be made: the single measurement of the components NO and NO_2 and the summary measurement of both components. Of these two possibilities the measurement of NO_2 is most problematic, because NO_2 is very easily soluble in water, leading to errors which cannot be checked.

Figure 3 shows a measuring installation for the determination of NO concentrations as well as of $(NO + NO_2)$ concentrations. The difference of the two measuring results already states the NO_2 proportion in the measuring gas. The metrological effort for a $(NO + NO_2)$ measurement is bigger on the extraction side, because in principle, heating the tubes and the accessories is necessary in order to prevent the high solubility of NO_2 in water.

To sample measuring gas, a sampling probe with a heated external ceramic filter is used. The connection line between sampling probe and analyzer is made of Teflon and stainless steel. The lines should be heated up in such a way that the line temperature is always 5 °C higher than the dew point temperature of the measuring gas. The infrared analyzer used is designed for the selective measurement of NO. Usually the measuring range is about 0–2000 ppm. For the determination of $(NO + NO_2)$ the measuring gas is forced through a converter by commutating switches. The converter consists of a heated silica tube with a controlled temperature of 1000 °C, effecting a quantative dissociation of the NO_2 into NO. The following cooler assures a constant dew point temperature of the measuring gas, about +2 °C. The measuring gas is forced into the analyzer through a diaphragm pump, a membrane filter and a flow

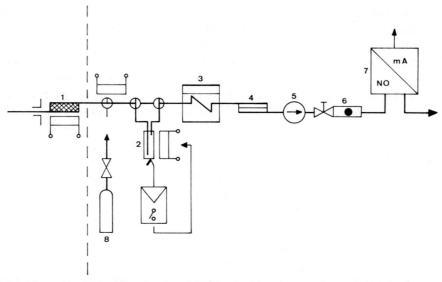

Fig. 3.   Measuring arrangement for NO$_x$-emission. 1 – sampling filter with heating device; 2 – converter; 3 – cooler; 4 – membrane filter; 5 – pump; 6 – flowmeter with needle valve;
7 – analyzer for NO; 8 – test gas.

meter. The zero point and end-point adjustment of the analyzer is made with air or
test gas. The test gases are always fed in before the cooler, in order to assure the same
dew point for testing gas and measuring gas.

1.3. MEASURING EMISSION OF VEHICLES

The emission of harmful substances by vehicles and its limitation is the most important air pollution problem. Investigations have shown that about 40% of air pollution
in towns is due to the exhaust gases of vehicles. The main injurious components of the
exhaust gases are carbon monoxide (CO), hydrocarbons (CH) and nitric oxides (NO
and NO$_2$). The harmful substances are mainly emitted by the exhaust and by the
crank case of the engine. Hereby the emission amount is rising in the rate of the
fuel/air throughput; in view of air pollution vehicles with a small cubic capacity and
low revolution number are to be preferred. Carbon monoxide and hydrocarbons arise
mainly from insufficient combustion, i.e., to diminish the CO and CH emission, and
to aim at an optimum combustion or an optimum ratio fuel/air. But on the other hand
we have seen that the emission of nitric oxides is highest for an optimal combustion.
As far as possible combustion should occur at a point where the minimal amount of
harmful substances are emitted.

The problem of air pollution caused by vehicles was first discerned and systematically
investigated in California, due to the enormous number of motor vehicles and the
extremely unfavourable weather conditions. A method of testing resulted from this
investigation, the so-called California test: using this the average amount of exhaust
gases and the harmful substances contained can be determined. This California test

and the modified Europe test – for European conditions – are meanwhile basic for the regulations of automobile exhaust gases for almost all countries.

As the composition of the exhaust gases differs according to the operating method of the engine, the California test considers a representative driving program including acceleration, steady drive, deceleration, and idle running, corresponding essentially to a drive through a big town, referred to as Los Angeles (Figure 4). The driving

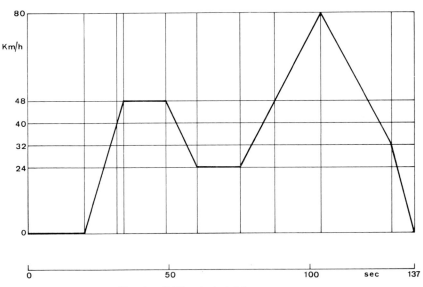

Fig. 4.   California test drive program.

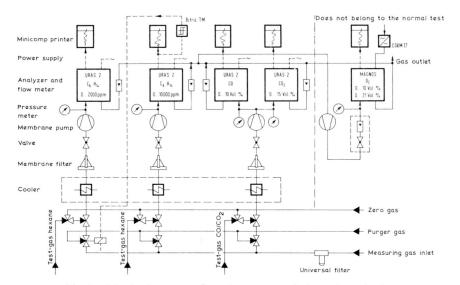

Fig. 5.   Monitoring system for exhaust gas analysis – schematic view.

program is effectuated on a dynamometer test station, compensating the different masses of car types by variation of the dynamometrical brake power. During the whole test, the concentrations of carbon monoxide, carbon oxide and hydrocarbons (measured as hexane) and eventually of nitric oxides is determined by a specially designed measuring installation (Figure 5).

As the single driving states within the test program are relatively short, very low reaction times of about 2 or 3 seconds – measured on the total installation – are required for the dynamic analysis of exhaust gases; this can only be obtained by an optimum design and high gas throughputs. As the throughputs are about ten times as high as in normal analysis installations, the accessories needed for the gas sample preparation and transportation have to be designed accordingly.

Fig. 6.   Monitoring system for exhaust gas analysis.

A sampling probe extracts the gas sample from the exhaust of the automobile to be tested; the sample is cleaned of dust by the input filter and condensed water in the gas line. A special liquid cooler guarantees a constant dew point temperature of the measuring gas under all working conditions. The condensed water is automatically drained by magnetic valves after each testing period. Before the cooler, there is a magnetic valve block for quick change from 'operation' to 'test gas', because the analyzer has to be recalibrated after each test program, according to the regulations.

The different components can be determined selectively and continuously by separate infrared analyzers. The analysis branches are mounted in parallel for the single measuring components in order to assure a synchronic response of the analyzers. For the measurement of hydrocarbons two analyzers which are calibrated for hexane and equipped with different measuring ranges are used. The analyzer with the largest measuring range is constantly switched in the measuring gas flow, while the analyzer with the smallest measuring range is only switched on automatically into the measuring gas flow for cencentrations lying within its measuring range; during the other time it is purified with air. These measures should prevent the Hexane analyzer with the small measuring range showing a time lag, caused by blowing up the high Hexane content in cases of quick changes in gas concentrations from high to low ranges. An additional measuring apparatus for oxygen existing for some installations are only good for a few general engine tests and they do not belong to the California test.

The complete analysis apparatus, including the gas preparation system and the recording instruments, is designed as a compact, movable unit and can therefore be directly used at the dynamometer test station (Figure 6). Besides the usual evaluations of the recorder strip charts, an automatic evaluation can be effectuated by a connected computer.

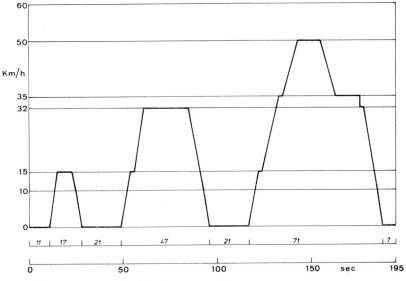

Fig. 7.   Europe test drive program.

The different traffic conditions in Europe, compared to those in the United States, need an exhaust gas test modified for the European situation, taking into consideration the average driving attitude in European cities. In comparison with the California test it displays considerable differences in the idling phase and in the maximum speed (Figure 7). The Europe test, resulting from these aspects, still has an even more important difference: the exhaust gas concentrations are not determined continuously during the driving program, but as an integral average value by means of a sampling out of a collecting bag for exhaust gases.

In consideration of the measuring techniques the same analysis installations can be used both for the California test as for the Europe test. The analysis apparatus for the Europe test can be essentially simplified because the Europe test does not have special requirements for the time response.

While hydrocarbon percentages were measured as representative concentration as Hexane with infrared analyzers up to now, today flame ionization detectors are used to determine the amount of hydrocarbons.

A new testing method – the CVS test – first adapted to the American conditions, does not directly evaluate the exhaust gas concentration, but analyzes the concentration of the real harmful substances within a certain exhaust gas – air mixture following the real environment conditions. Here the total car exhaust gases are directly analyzed continuously after rarefaction with an eight-fold quantity of fresh air, or its integral mean value is determined, having been collected in a sampling bag. This method has the advantage of preventing additional reactions due to temperature, and solubility errors due to condensed water, because of the rarefaction of the exhaust gas directly behind the escape. On the other hand there are higher requirements for the analysis technique, because the measuring ranges of the analyzers must be designed ten times more sensitive than in former test methods, due to the high rarefaction of the exhaust gases.

The analysis apparatus for the CVS test has a flame ionization detector to determine the total amount of hydrocarbons and two infrared analyzers to determine the contents of carbon monoxide and carbon dioxide. For the additional nitric oxide analysis a new measuring method is prescribed; it is based on the chemo-luminiscence reaction between ozone and nitrogen monoxide.

Since the above-mentioned testing methods are principally used for the type-testing of new car models, there is a decree subsisting for the Federal Republic of Germany proscribing an exhaust gas control for carbon monoxide; this control is effectuated in the course of the technical control which all licensed cars have to undergo every two years. The CO-emission must not exceed 4.5 vol % for a warm engine when idling. According to the required selectivity and precision a license could only be granted for adequately designed infrared analyzers, while other measuring methods have been rejected. Special portable analyzers based on infrared absorption were designed to analyze the exhaust gas for carbon monoxide in idle running; these instruments equipped with built-in sampling apparatus allow an ambulant application directly on the vehicle.

1.4. IMMISSION CONTROL BY MEASURING STATIONS AND MOBILE MEASURING STATIONS

In order to recognize level and development of the air pollution and to take suitable corrective measures a conception must be found for suitable measuring stations, in order to measure the immission values of air pollutions. The Federal Government of Germany, for example, has already enacted a law on "precautions against air pollution" in 1965, according to which measurements concerning kind and quantity of air pollution by gas and dust and measurements concerning the meteorological conditions are effectuated to prepare and to execute federal regulations.

For the design of ambient air control stations it should be considered that the meteorological conditions have a major influence on the fluctuations of the immission at one point, so it is absolutely essential that they have to be determined. Because of the major and rapid changes in the concentration of air pollution as well as the meteorological values in connection with the requirement for economic evaluation of the data automatic apparatus continuously in operation has to be used. Only a continuous recording of measuring values can reveal whether present SO_2 concentrations result from a sulphuric acid plant or from combustion, since the CO and SO_2 content increases at a certain rate for combustion.

According to the above-mentioned law concerning air pollution, some large measuring stations were erected in Germany to determine the immission evaluations, and

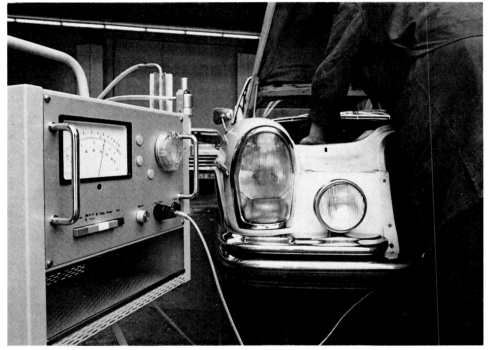

Fig. 8.   Exhaust gas analyzer GGCOT-2 in operation.

through these valuable experience could be obtained in the field of air control. The first test station was erected in Frankfurt on Main by order of the Federal Health Ministry. In Munich the local administration manages a large system of stations; their measuring values are centrally digested by a balance computer. Another station network was erected in Berlin under a research order of the Federal Health Department. Besides this, a number of industrial enterprises also maintain measuring stations specially designed for the pollutions emitted by these respective industries. To explain the construction of such a stationary measuring station for the determination of immission values, the Frankfurt Station will be treated in the following.

The measuring station consists essentially of a main station in the city centre and two sub-stations on the Eastern and Western sides of the city (Figure 9). Sulphuric dioxide, nitric dioxide, carbon monoxide, carbon dioxide, hydrocarbons, relative humidity, dust, radiation balance, temperature, wind direction and wind velocity are measured in the main station. Sulphuric dioxide, hydrogen sulphide, nitric dioxide and carbon dioxide are measured in both sub-stations. The choice of the gaseous measuring components depends essentially on the branches of industry in the controlled region.

To measure the SO_2 concentrations, a gas trace analyzer PICOFLUX is used, which can also be used for NO_2 measurements in a modified model with an electrochemical measuring cell. The components CO and CO_2 are controlled with seperate infrared analyzers URAS; the CO analyzer is a special model with a prolonged measuring cuvette for small measuring ranges. The infrared analyzers are recalibrated twice a day by an automatically working balancing instrument. A flame ionization

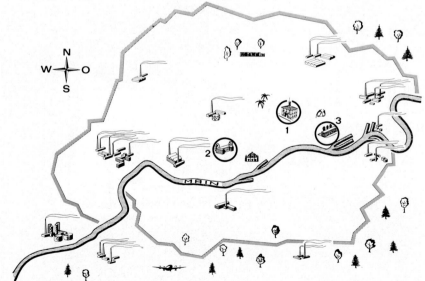

Fig. 9.   Location of the measuring stations in Frankfurt/Main. 1 – Main station in the city; 2 –
Substation in western direction; 3 – Substation in eastern direction.

detector is used for the determination of the hydrocarbons, which respond to the total of the organically bound hydrocarbons. Dust concentrations are measured by optical means; the dust particles are deposited on a filter tape, the light transmission of which is scanned by means of a photocell, and wind direction and wind velocity are determined by a vane with a resistance teletransmitter or a wind sensor with accessory turbine. Temperature and air humidity are easily determined by a resistance thermometer or by a hair hygrometer with resistance teletransmitter. The measuring of the radiation balance is effectuated by thermo-electrical method by means of multicellular thermocouples.

All the measuring values are continuously traced on pen recorders in order to find out the emitters by comparing the transient values. But for the total evaluation an average value over half an hour is needed. For this purpose a balance computer is used; this is fully automatic because the usual evaluation methods, for example, planimetering of the strip chart, were not suitable due to the high number of measuring values.

While the signals of the main station can be fed directly into the balance computer, the signals of the sub-stations have to be transmitted to the main station via telephone cables. For a better control of the single measuring instruments the measured values of the sub-stations are additionally recorded on a multipoint recorder in the main station. By means of the strip chart one can easily judge if the half-hour average values in the computer are correct.

Mobile measuring stations are quite good for complementing the stationary measuring systems; they can be used near already existing stationary stations, when these are not sufficient in case of danger. At the same time, control can be realized fairly economically with a mobile station, where the permanent construction of permanent measuring stations would not pay due to the relatively low degree of air pollution. Besides this it would be a good idea to carry out a check with a mobile station before constructing a stationary measuring station system, to see whether the planned locations the for stationary measuring stations were well chosen. The technical conception of such a mobile station is described in the following.

A standard lorry chassis with a special mounting is used as the mobile station (Figure 10). The height of the special mounting allows upright walking inside the vehicle. The mobile station is additionally divided into two parts: the driver's cab and the measuring cab. The measuring cab is also divided into two parts by a measuring panel. So there is a room behind the measuring panel, which is accessible from the outside by a separate door. Here are all the cables, the stock bottles for the test gas and the pneumatic measuring gas lines. In front of the measuring panel we find the actual workroom (Figure 11). The whole car is air-conditioned in order to guarantee constant working conditions. The analyzers, the recording instruments and the electric indicating instruments for the mains supply of the vehicle are built into the measuring pannel. On the left and on the right of the measuring pannel are the test racks, in which is the calibrating installation for the analyzers, together with refrigerators, the dust monitor and the computer with typewriter and perforated strip punch. The elec-

Fig. 10.   Mobile measuring station for air pollution.

trical supply of the mobile station is usually effectuated by cables from the outside. Additionally an emergency mains supply can be provided by means of a built-in battery set. An emergency current aggregate for the operating of the vehicle has not been built in, because it could initiate other problems concerning environment protection and falsification of the measured data. On the rear surface of the vehicle a crank pole for mounting a wind direction and wind velocity detector is fixed. The sampling line for measuring gas can be fixed to the crank pole, in cases where the concentration measurements have to be carried out at different heights.

The metrological equipment of the mobile station mainly corresponds to the equipment of the stationary measuring stations mentioned above. Newly developed gas trace analyzers PICOS – as an insert unit – are used for the determination of the nitric oxides and to determine the concentrations. A gas mixing device is used to control the trace analyzers, allowing a mixing rate of maximum 1:1000. In this way concentrations can be analyzed in the ppb range by using premixtures.

All the detectors for concentration values and meteorological variables give an standard d.c. output signal of 0–20 mA. These signals are transformed into a current proportional number of pulses by an electronic pulse detector and they are integrated over a period of three minutes. These three-minute average values are printed and punched on a teleprinter together with date and time. The punched tape enables one

Fig. 11.   Interior view of the mobile measuring station for air pollution.

to transmit the dates to a computer. Based on the concentration values in connection with the meteorological data, this computer can give classifications, frequency distributions, or quotient constitutions, from which a conclusion concerning the pollution emitters can be drawn.

1.5. AMBIENT AIR CONTROL

Ambient air control concerning harmful gas pollutions is very important, because those working in these stations can be projected against health injuries. This is very important if you take into consideration the fact that a person inhales about ten to twenty cubic meters of air comprising all kind of gases and vapours during an eight-hour work day. Considering the high absorption capacity of the lungs, a great many toxines can be absorbed into the organism. By judging the injurious effect to health of the different gases and vapours, one cannot give a global classification into injurious and non-injurious gases. Every gas, even oxygen for example, can be injurious and even fatal at a high enough dosage. The injuriousness of a gas depends on its duration of effect as well as its concentration. For ambient air control the toxic limit, the concentration which would surely lead to death within a short time, is not so much of interest, as a concentration which would surely not do any harm even for a long period of time. The allowed maximum concentration at the place of work, referring to an eight-hour effect per day is called the "maximum working place concentration", or the 'MAK'-value in Germany. A summary of the MAK values for the different injurious gases and vapours is published by the trade associations and is continually brought up to date by the latest research data. In other countries there are comparable critical values, which partly differ from the German MAK values in their definition and single values. The most well-known gas, regarding its injurious effect, is probably carbon monoxide (CO). Carbon monoxide is mostly to be found in the exhaust gases of motor vehicles, sewer gas, water gases, coke plant gases, and generator gases as well as in all sorts of kilns and in the chemical industry. By means of a diagram of the toxic effect of carbon monoxide we see that even low CO concentrations can provoke toxic symptoms for a sufficiently long duration of effect (Figure 12). In consideration of these conditions, the MAK value for CO is fixed at 50 ppm.

Ambient air control consists of measuring the concentration of the injurious gas and in signalling the exceeding of the maximum permissible concentration. The indication and the signalization of the measured values has to occur so quickly, that the endangered persons can be averted in time and that corresponding counter-measures such as the switching on of ventilators or switching off of throttling machinery can be

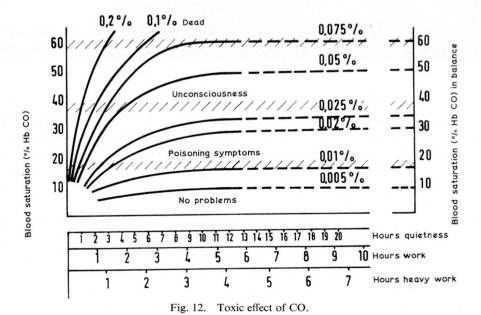

Fig. 12.   Toxic effect of CO.

initiated. In certain automatic installations the starting device for the counter-measures can be directly linked to the measuring instrument. As the gases do not diffuse over the whole room at once, it is not sufficient to have only one measuring point. For this reason several measuring points are usually located in one room; they are periodically controlled by an analyzer, since linking an analyzer to every measuring point would be much too expensive.

The local distribution of the measuring points in the room is of decisive importance. The gas sampling devices should be arranged in such a way that measurements are realized mainly at points where people are working and not so much at points where the injurious gases and vapours mainly occur. The periodic scanning of the measuring points is made by a program-controlled solenoid valve stock or by an automatic gas commutator switch (Figure 13). One needle valve corresponds to every sampling line,

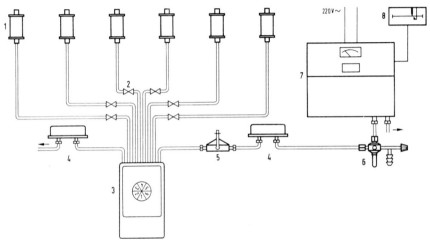

Fig. 13.   Measuring arrangement for ambient air control. 1 – sampling filter; 2 – needle valve; 3 – gas selector switch; 4 – pump; 5 – membrane filter; 6 – flow meter with needle valve; 7 – gas analyzer; 8 – indicator.

the single gas lines are adjusted to the same pressure drop. The sampling lines, which are not connected to the analyzer for the time being are pre-sucked off in order to improve the time response, so that there is always fresh measuring gas at the analyzer, when switching over to the next measuring point. The response time for each measuring point results from the number of measuring points, the length of the lines, the pump power and the response time of the analyzer. Usual values are about $\frac{1}{2}$ to 2 minutes. The choice of the analyzer or of the measuring method depends on the controlled gas components. The measuring range of the analyzer is always larger than the respective MAK value in order to determine short period excesses over the limit value and to analyze them.

The signal circuits are usually so designed that an acoustic warning signal is released by means of the contacts of an indicating controller and at the same time signal lamps attached to the measuring points flash when the adjusted limit value is exceeded.

Moreover a fan can automatically be switched on by the same contacts a relay. The measuring points of the signal are synchronised by pneumatic gas commutator switches. As far as an additional recording of the measured values is desired, a multi-point recorder with remote controlled commutator switch and remote controlled print-out is used, ensuring an additional measuring point identification of the recorded traces.

2. The Measuring Methods

2.1. GENERALITIES

In the field of industrial measuring techniques the analysis of gas mixtures with con-tinuously working measuring instruments is getting more and more important. Com-pared to the discontinuous methods of chemical hand analysis the continuously mea-sured data is quite an advantage, because only continuous and automatic functioning ensures complete information on changes in concentrations. In this context 'analysis' always means the quantitative determination of the concentration of one component of gas mixture; the qualitative composition of the gas mixture must be known.

The basic problem of each gas analysis is the continuous determination of the con-centration of one component, the so-called measuring component in a multi-compo-nent gas mixture. Hence it follows that the measuring method has to be selective for the measuring components, i.e., advantage should be taken of the physical and chem-ical properties of the different gases, where the measuring components differ clearly and if possible specifically from the accompanying components. It is also presumed that a reliable and serviceable measuring method has to be found consequent with the relevant physical and chemical properties. If possible, analyzers working on a purely physical measuring principle should be used because an automatic operation hardly needs any servicing, at the same time of strong construction. If the measuring sensi-tivity of the physical method is not sufficient, chemical and chemico-physical mea-suring methods should be applied, enabling one to measure even the lowest gas con-centrations.

2.2. INFRARED GAS ANALYZED

Among the known continuous analysis methods the infrared measuring principle has undoubtedly the greatest field of application due to the large number of measurable components and to the practicable measuring ranges. The infrared method is applied for almost all emission measuring problems as well as for some components such as carbon monoxide and carbon dioxide, and also for immission measurements in the field of air pollution control. The measuring effect of the infrared analyzer is based upon the fact that all heteroatomic gases absorb electromagnetic radiation in pro-nounced wavelength ranges, the so-called bands, which are specific for each gas (Fig-ure 14). This measuring method is not suitable for gases having no electrical dipole moment, such as the monoatomic rare gases as well as the equal atomic elementary gases such as nitrogen, oxygen and hydrogen.

The operating range of the infrared analyzer URAS 2 T is of 2.5 to 12 μ wavelength

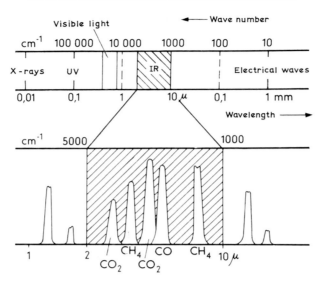

Fig. 14.   Electromagnetic radiation spectrum.

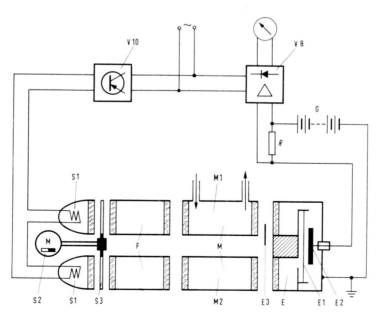

Fig. 15.   Schematic view of the infrared gas analyzer URAS 2 T. E – receiver; E1 – membrane; E2 – counter-electrode; E3 – receiver restriction; F – filter cuvette; G – D.C. voltage source; M – measuring cuvette; M1 – analysis chamber; M2 – reference chamber; R – high-impedance resistor; S1 – radiation sources; S2 – diaphragm edge motor; S3 – diaphragm edge; V8 – amplifier; V10 – mains stabilization.

in the infrared spectral range. The measuring of the absorption is effectuated by a non-dispersive method, i.e., there is no spectral dispersion of the infrared radiation. The absorption is measured in an altering light photometer arrangement with two parallel beam paths and a selective radiation receiver (Figure 15). Two identical infrared emitters which are heated in a defined manner, the radiation of which is modulated in phase by a motor-driven diaphragm wheel, are used as a radiation source. An hermetically sealed chamber, which is divided into two parts by a metal membrane is used as a radiation receiver. The radiation receiver is always filled with the gas whose concentration content is to be determined. Between radiator and receiver there are two separate cuvettes which are equipped with windows permeable to infrared radiation. The infrared radiation comes into both receiver chambers with the same intensity if both cuvettes are filled with a gas which does not absorb any infrared radiation, for example nitrogen. The radiation can only be absorbed in the range of the bands which are specific for gas filling the receiver. The equal absorption for both chambers can be realized because of the interruption of the path of the measuring ray by the diaphragm edge in periodical temperature and pressure fluctuations, which are equal in phase and intensity. Thus the pressure fluctuations do not take effect on the metal membrane between the two chambers. If the gas mixture that has to be analyzed is passed through one of the cuvettes, the so-called measuring cuvette, while the reference cuvette is filled with a gas that does not absorb radiation, the path of the ray is more or less attenuated according to the concentration of the measuring component; this implies a different absorption in both receiver chambers. This again causes differences

Fig. 16.   Infrared gas analyzer URAS 2 T.

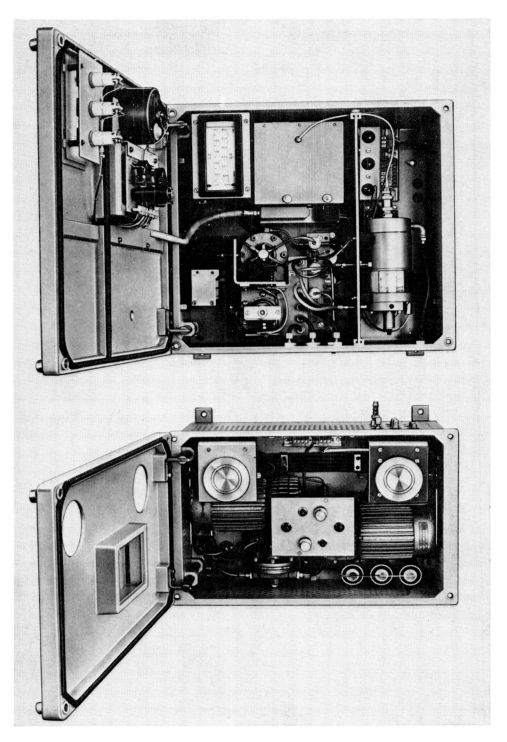

Fig. 20.   Flame ionization detector FIDAS.

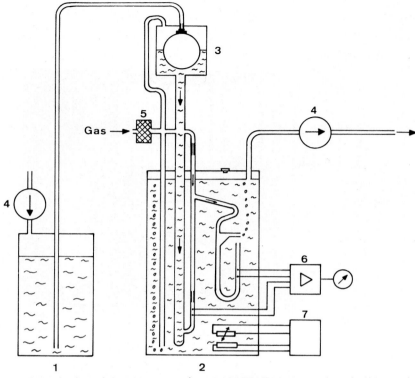

Fig. 21.   Schematic view of the gas trace analyzer PICOFLUX 2. 1 – stock tank with reaction solution; 2 – measuring unit; 3 – level regulator; 4 – pump; 5 – selective filter; 6 – amplifier; 7 – thermostat.

centration of the measuring component. Weakly acidic distilled water with an addition of hydrogen peroxide, which is quantitatively converted into sulphuric acid by reacting with the sulphur dioxide contained in the measuring gas, is used as a reaction solution.

The conductivity is determined by means of a reference cell before the reaction and of a measuring cell after the reaction (Figure 21). The difference of the two conductivities produces an alternating current which is proportional to the concentration of the components; an amplifier supplies a 20 mA d.c. current standard output signal. From an external supply vessel the reaction solution is forced to a level regulator by means of a diaphragm pump, which is built into the analyzer; the regulator guarantees a constant input pressure of the reaction solution. It assures a constant flow of the solution by means of a calibrated calibrated capillary. The measuring gas is dosed by a capillary by means of a second diaphragm pump (suction pump), mounted in the output of the measuring system. The constant input pressure of the measuring gas is maintained by means of an immersion tube. The measuring system is located in a thermostated liquid bath in order to avoid ambient temperature fluctuations of measuring and reaction solution. Measuring gas and wasted reaction solution are both sucked up by the liquid bath serving as immersion vessel for the necessary measuring gas input pressure.

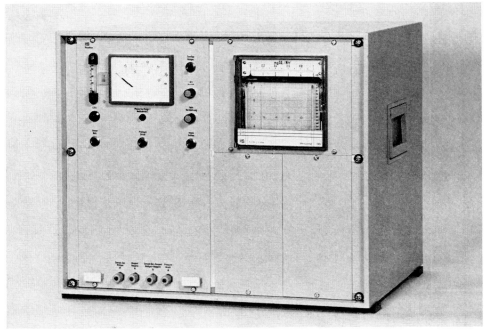

Fig. 22.   Gas trace analyzer PICOFLUX 2.

The PICOFLUX 2 is designed as an insert unit instrument in the 19″ technique (Figure 22). There are always two standard measuring ranges of 0 – 1 and 0–5 mg SO_2 m^{-3} with automatic measuring range commutation. A reaction solution supply of 10 l ensures a continuous operation of at least one month without any servicing: this complies with the requirements for the application in stationary and mobile measuring stations.

The gas trace analyzer PICOS is a novel measuring instrument. The PICOS is based on an electro-chemical measuring method, which is very apt for the immission measurements of toxic gases not often found in air, due to its high measuring sensitivity. The ion equilibrium of the electro-chemical system in the measuring cell is disturbed by the reaction of the measuring gas with a specific electrolyte. A current, which is rigorously proportional to the concentration of the measuring component, flows in the external circuit. Contrary to the known methods a new organic solid electrolyte is used instead of a liquid electrolyte. This solid electrolyte, containing all reactants in sufficient quantities, enables one to design very compact measuring cells, in order to obtain small and handy instruments.

A diaphragm pump sucks the measuring gas and forces it through the cell via protective and selective filters (Figure 23). As the current supplied by the measuring cell is proportional to the mass flow of the measuring component, a driving pump with a constant stroke is used, the frequency of which is controlled as a function of the temperature. The same mass of the measuring components reacts per time unit in the

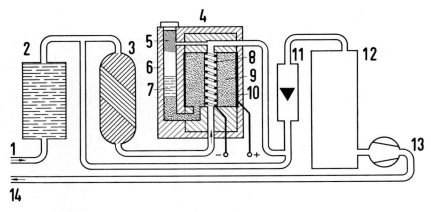

Fig. 23. Schematic view of the gas trace analyzer PICOS. 1 – gas inlet; 2 – protective filter; 3 – selective filter; 4 – measuring cell; 5 – filter; 6 – carrier block; 7 – regenerator; 8 – measuring electrode; 9 – electrolyte; 10 – counter electrode; 11 – flow meter; 12 – buffer volume; 13 – pump; 14 – gas discharge.

Fig. 24.   Gas trace analyzer PICOS.

measuring cell for equal volume concentration. The chemical reaction and the electric charge exchange within the measuring cell occur at the threephase boundary layer: measuring gas– electrolyte– electrode. The choice of the electrolyte and of the electrodes is very important for the selectivity of the measuring arrangement for certain measuring components. Disturbing accompanying components can be additionally eliminated by means of suitable selective filters.

The gas trace analyzer PICOS is designed as a compact measuring instrument, based on the principle of the unit composed system. An independent operation without external accessories is possible. A 1/4 19″ insert unit comprises a protective and selective filter, a measuring cell, flow meter, gas suction pump, amplifier and indicator as well as power supply with charging device for the builtin accumulator (Figure 24). The electric function units are mostly designed as plug-in cards, exactly like the pneumatic function units. Independent of the power supply continuous operation up to about twelve hours is possible by means of a built-in accumulator.

The analyzer is first of all designed for the measurement of hydrogen sulphide and nitric oxides in immission measurements. By developing suitable electrolytes and electrodes in the future, further components can be determined. The possible measuring ranges are between a few ppb and a few ppm, depending on the measuring components and on the composition of the measuring gas.

RECENT ADVANCES IN AUTOMATIC CONTINUOUS CHEMICAL ANALYSIS, MONITORING AND CONTROL

H. C. HELLIGE

DELTA Scientific Corporation, Lindenhurst, N.Y., U.S.A.

Abstract. The development of a unique continuous, automatic wet chemistry analyzer, which exhibits long term test accuracy and freedom from serving is discussed. The special design features that overcome deficiencies in previously available commercial analyzers which use capillary flow systems or peristaltic pumps, are also indicated.

The history of the automation of wet chemical analyses undoubtly goes back to the earliest chemical experimenters. It is obvious that many tests which have to be endlessly repeated, such as the continual determination of water hardness, or the determination of the concentration of treatment chemicals in industrial process waters for ammonia, phosphate, silica, hydrazine, chlorine, etc. are ideal candidates for automation. More recently, with the tremendous acceleration in interest to detect and control pollution, there has been a vast increase in the need to automatically monitor process streams and waste effluents in scores of diverse industrial and commercial operations such as electroplating, pulp and paper, chemical manufacturing, food processing and the like.

Many widely-used and historically-proven wet chemical tests make use of chemical reagents which are added in measured amounts to a specific quantity of the water sample, and which consequently produce a colorimetric reaction. With the refinement of photoelectric colorimeters to accurately measure these resultant colors and compare them electrically to the colors produced by known standard values of the substance under test, a significant 'automatable' improvement became available to substitute for the color-comparing eyes of the chemist.

It appeared that all that remained to be done, was to substitute automatic methods of measuring sample quantities, and to sequentially measure and add the reagents required in the test. In actual commercial practice, this seemingly simple step presented by far a much greater challange to man's ingenuity than the automation of the color-comparing eye.

Two fundamental methods were originally devised to perform the sample-treatment and reagent-addition steps automatically; and in most commercial analyzers, variations of these older methods are still in use today, in spite of severe limitations they pose on accuracy and service-free operation.

One method attempts to control the flow of sample and reagents by means of glass capillaries together with overflow devices and 'constant reagent water-head' devices which present a constant pressure of fluid to the restriction of the capillary. As can easily be imagined, the flow in the capillaries is subject to easy clogging, to gradual errors and to service problems due to the build-up of foreign matter in the fine capil-

G. Lindner and K. Nyberg (eds.), Environmental Engineering, 189–115. *All Rights Reserved*

laries, and to errors due to variations in pressure-head as the reagents are consumed.

To overcome the problems of capillary feeding, analyzers have been offered by various manufacturers in which the sample and reagent-measuring steps are performed by a variety of pumps. Two fundamental types of pumps have been offered prior to the development of the Delta Scientific system, and each of the older pump methods suffer from their own deficiencies. Piston or rotating-cylinder or cone pumps, which are designed to capture a specific, exact quantity of sample or reagents, suffer from leakage and lack of true percision under continual use. Many chemical tests make use of reagents which are sensitive to most gland-packing materials, and attempts to provide critical pump parts that were both non-contaminating as well as very long-wearing have been fruitless. Another type of pump widely offered is the conventional perastalltic type in which flexible plastic tubing is squeezed shut in order to trap and move forward a segment of sample or reagent. This type of pump suffers severely because the tubing, under such severe service, must be replaced at frequent intervals... with constant need to recalibrate the entire device. A further disadvantage of the perastalltic pump is that there is no way to easily adjust the exact quantity of reagent being metered, since the tube must be totally squeezed shut and only gross quantities of reagents are able to be trapped as the squeezing action presses the fluid on. This deficiency is attempted to be overcome by careful selection and matching of the plastic tubing sizes, but this itself imposes a severe limitation on exact-volume selectibility, and in addition poses an even greater service problem because of the perastalltic pump's frequent need for tubing replacement.

1. Development of a Superior Pumping and Metering System

A careful examination of these inherent problems with older pumping methods, has lead the research and development division of our company to devise an unusually unique pumping and metering system, which completely overcomes these limitations. All details cannot be explained at this time because of pending patent considerations. However, in general outline it can be said that the unique Delta Scientific Series 8000 Pumping and Metering System consists of a series of parallel plastic tubing channels (in most cases six channels), each of which has a unique set of inert plastic, one-way flap-valves, with a valve below, as well as above, a central 'stroking' section. A specific, exact volume of sample or reagent is 'captured' at all times in this section between each of the two valve sets (Figure 1).

To perform the required pumping action, a motor-driven six-segment cam simultaneously rotates against six hinged pressure springs, each of which is in flat contact with one of the six 'captured-fluid' flexible plastic tubing sections. As the cam rotates, its action gently strokes each of the six pressure springs, *partially* squeezing in on the 'captured volume' of fluid in each tubing. The upper valves opens automatically to relieve the built-up pressure in each tubing, and upon the cam's further rotation the hinged pressure springs withdraw their pressure on the six tubings. The resiliency of each tubing returning to its normal unflexed shape, creates a negative pressure, and

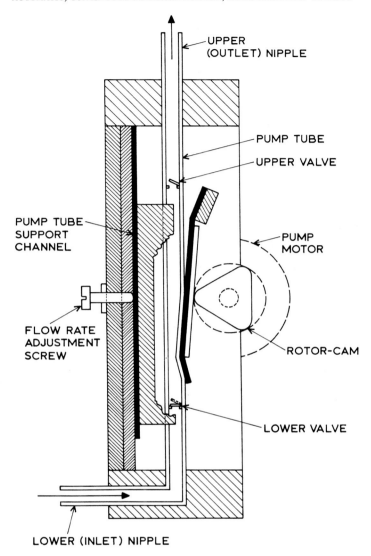

UPPER
(OUTLET) NIPPLE

PUMP TUBE

UPPER VALVE

PUMP TUBE
SUPPORT
CHANNEL

PUMP
MOTOR

FLOW RATE
ADJUSTMENT
SCREW

ROTOR-CAM

LOWER VALVE

LOWER (INLET) NIPPLE

Fig. 1.   Delta Scientific series 8000 multi-channel metering pump for
automatic continuous wet-chemistry analyzer.

the bottom valve in each segment automatically opens, drawing further fluid into the
the 'captured segment' to replace the exact amount that has been pumped upward into
the analyzing section on the previous stroke of the cam. This unique design completely
avoids the total distortion of the plastic tubing required by perastalltic pumps. Its
gentle stroking action completely eliminates the problem of frequent tubing replace-
ments, because at no time is the distortion of the tubing so great as to cause unusual
stress or wear on it. Thus, one of the two major disadvantages of the perastalltic

Fig. 2.   Delta Scientific series 8000 continuous automatic wet-chemistry analyzer employing simultaneous metering of sample and blank streams with four different reagents. Indicating meter has high and low set points for alarm and relay-control of treatment system valves.

Fig. 3. Delta Scientific series 8000 continuous automatic wet-chemistry analyzer with digital test-result indication and BCD signal output for computer control of process variable. Test requires but one reagent, however, multi-channel pump simultaneously feeds test sample, dilution water and blank sample for automatic elimination of the effects of original colour and turbidity in the sample.

pumping method – the wear and service problem – has been completely solved by the new Delta Scientific Series 8000 Pumping System.

The second, most serious defect of the other pumping systems – the inability to finely adjust the precise volume of the various streams, has also been completely solved in the Delta Scientific Series 8000 Pump. Opposite each pressure spring, on the other side of each of the six tubing sections, is an *individually adjustable* tubing-support channel. A threaded screw, which can be advanced or withdrawn, permits the entire tubing section between the valves to be positioned as desired – closer or further away – from its respective pressure spring. This unique ability to adjust the depth of penetration of the pressure spring in each of the different channels, makes possible an exact adjustment of the desired volume to be metered by each channel. This technical breakthrough, because of its total adjustability, furthermore makes it possible to use ordinary, unselected plastic tubing for the pumping channels. It eliminates the need to carefully measure and match tubing sizes, and makes unnecessary the time-consuming trial-and-error technique of calibrating perastalltic-type pumps, because each channel's exact metering volume can instantly and easily be infinitely adjusted to meter more or less fluid by simply turning the screw adjustment.

Each Delta Scientific Wet-chemistry Analyzer/Controller is a self-sufficient system, uniquely assembled for the user's exact needs. Whether housed in modules or in a single case, each consists essentially of a system for treating the fluid sample stream with reagents and a system for making simultaneous colorimetric measurements of the treated and untreated sample streams.

Delta's method of making colorimetric measurements is unique in that interference due to turbidity or color naturally present in the liquid is automatically overcome, so that the test results represent the exact concentration of the species present. In applications where turbidity or color interferences are known to be present, or where they may occur, the sample stream is split so that a portion flows through each of two sample cells. The stream which is to be treated with reagents is termed 'sample' or 'test' stream. The second, which is not to be treated, so that any inherent turbidity or natural color in it can be electrically subtracted from the total measurement, is called the 'reference' stream. Both the 'sample' and the 'reference' circuits use a common light source and incorporate matched photo-optical systems. Glass color filters of pre-determined wavelengths are mounted in the light paths of the two flow cells. The optical color filters are specially designed for each Analyzer based on the range of test, for highest possible sensitivity. The difference in electrical output signals is measured and, since the reference circuit eliminates the effect of interferences, the net signal is proportional to the amount of the species present.

Fluid for the photometer sample cell and reference cell is drawn from a common supply and is delivered at equal rates over equal time periods to the respective flow cells. This assures a precise correlation between the reference sample and the treated sample. In certain special circumstances, such as the direct measurement of the inherent color of a sample stream, a separate reference stream is not utilized. In these cases, a modified split-beam photometer is used to make two simultaneous measurements on the sample stream. Delta's unique, multi-section pump simultaneously meters sample streams, reference streams, reagents and dilution fluids if needed.

The electrical signal from the photometer is amplified and is then indicated on the meter of the Analyzer. A separate output signal is available for recording or for proportional control of valves, pumps, etc. Digital output for computer interfacing and other signal levels for telemetry has also been provided. Each analyzer system has as standard equipment 'high-low' set points on the meter, and associated relays for on/off control of valves, pumps, etc., and for remote alarm or signaling purposes.

Successful application of Delta Scientific's Series 8000 Wet Chemistry Automatic Analyzers have been made in a variety of industrial, municipal, and commercial applications. Continuous Analyzers are in use for such diverse tests as ammonia, chlorine, chromate, hardness, iron, nitrate, phosphate, silica, sulfite, colour and hydrazine, and many different additional tests are now undergoing testing. Typical units are shown in Figures 2–3.

PART V

WATER POLLUTION

General

CHEMICAL ENGINEERING ASPECTS OF WATER POLLUTION CONTROL

ROBERTO PASSINO and APPIO C. DI PINTO

Instituto de recerca sulle acque and University of Rome, Rome, Italy

Abstract. In this paper the importance of a study of water treatment processes based on the principles of classical chemical engineering is stressed. Biological reaction engineering plays an increasing role in defining new necessary unit operations. The importance of inplant measures to prevent pollution rather than abate it is pointed out.

A certain amount of difficulty was encountered by us in constricting such a vast and complex subject into such narrow limits.

As a result, only some of the aspects of chemical engineering connected with water pollution control have been examined.

The points which we have tried to emphasize are those of the role of the project engineer and the importance of a rational study of the whole subject based on the principles of classical chemical engineering.

1. General Remarks on Water Treatment Plants

A water treatment plant transforms a pollution influent into a treated effluent and a stream which is stabilized and concentrated in solids. Other streams may be constituted of chemical reagents, air, by-products, fuel, etc. ...

The degree of treatment of the effluent can vary according to the purposes for which the plant has been designed. It can be limited to obtaining an effluent which is acceptable with respect to the characteristics of the receiving stream and the uses for which the latter is destined, or on the other hand of treatment can be pushed towards obtaining an effluent with such characteristics that it may be re-used directly. The influent stream can be a municipal or industrial discharge, coming from one or more industries or urban communities, either separately or together. Among the principal reasons for the installation of a treatment plant the following may be listed:

(1) *Reduction of existing pollution.* The word 'reduction' is used instead of 'elimination' because it would be unrealistic to think it possible to eliminate pollution completely in an industrially developed society. An economic analysis of the benefits connected with complete purification would also show that these would not justify the cost required to construct and maintain the necessary plants.

(2) *Preventing of an increase in pollution.* The need, in this case, can derive from a foreseen increase in industrialization or in the resident or seasonal population in the zone served by the plant.

(3) *To allow the installation of industries in zones* which are not sufficiently endowed with water resources and where it is desired to promote economic and civil development.

G. Lindner and K. Nyberg (eds.), Environmental Engineering, 199–213. All Rights Reserved
Copyright © 1973 by D. Reidel Publishing Company, Dordrecht-Holland

A fairly eloquent example of this situation is the South of Italy. There, apart from the scarcity of water, the hydro-geological situation must be protected, also because of the development of tourism in the zone.

(4) Prevention of excessive discharges concentrated over a short period of time which could cause considerable damage to the aquatic environment.

(5) Maintenance of determinate ecological conditions in zones intended for recreation, tourism, bathing, fishing, etc.

(6) Other motives, such as: the increase of the total water resources available, which can be attained by means of the re-use of the effluents treated; allowing for the recovery of economically valuable by-products.

In any case, it is best that all these motives be analyzed, taking account of the water situation in the entire catchment area.

The importance, or rather the indispensability, of the treatment plant is evident; but before speaking directly about the various aspects of chemical engineering connected with the designing and the construction of the plants themselves, it is best to focus on a particularly important point which, from the logical point of view, takes precedence.

2. The Role of the Project Engineer in Water Pollution Control

In many countries, at least until a few years ago, the main if not the only preoccupation, of the designers of plants or industrial complexes, was essentially limited to the problems of production optimization, aiming at the manufacture of a product at minimum cost to the producer. In such a way the entire cost was determined only by the internal constraints of the production process, without any consideration of the so-called external technological diseconomies.

It is possible, however, that the installation of a certain industrial complex may cause the pollution of a water-source which would force the utilizers downstream to purify the water and to bear the cost of this. The total expense could then turn out to be greater then if these had been in the first plant, productive processes causing less pollution, and/or if the discharges had been subjected to treatment.

The resulting complex of these problems has sometimes been such that it ever compromises the ecological situation, destroying public goods and thus giving rise to an external diseconomy of vast proportions, which cannot easily be estimated in quantitative terms.

The avoidance or at least the reduction of pollution ought to be one of the essential factors for the designer the basis of the designer's project. The cost of pollution, calculated as the cost of installation and operation of the treatment plants necessary for the purification of the discharges must be taken into account in the same way as the cost of basic materials, energy, capital, etc. Therefore, the project engineer must also have an ecological frame of mind, such that, from the very moment of the designing, he may follow criteria which reduce to a minimum the polluting load, such as:

(1) When choosing the locality in which to install a plant, it is necessary to consider

the possiblities of integration with other activities in the same zone which allows, as far as possible, for utilization in a closed cycle of the potentially polluting materials.

(2) Developing new productive processes and modifying those already in use, in order to reduce the polluting output.

(3) Economically recovering the by-products. Examples of this kind are already

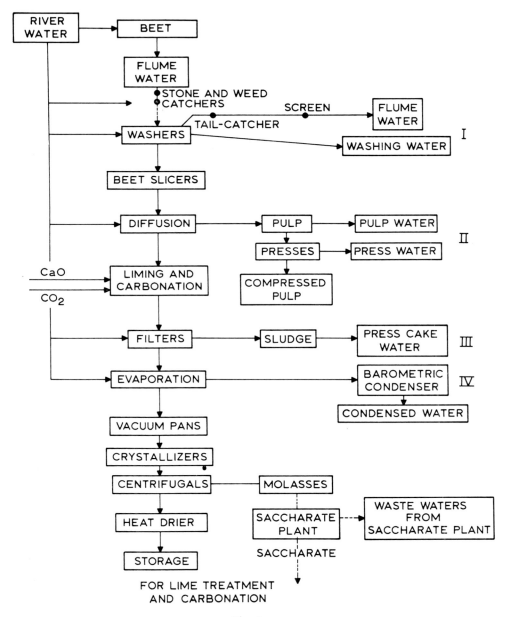

Fig. 1a.

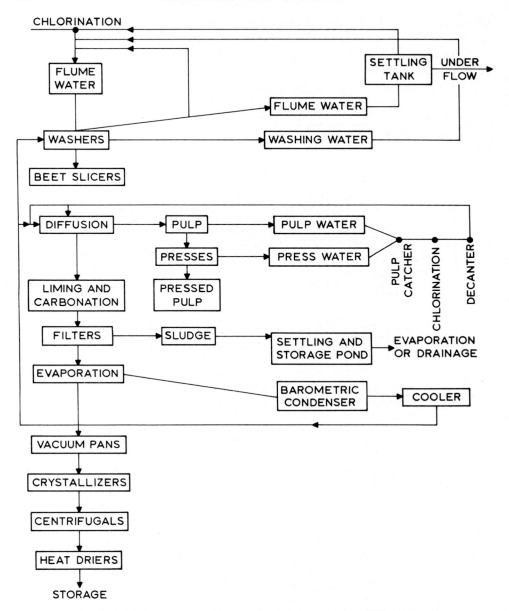

Figs. 1a–b. Water system with reuse, sugar industry. (From IUPAC, 1963.)

numerous and technical development renders these possibilities for recovery more and more accessible and frequent.

Separating the discharge collecting networks of industrial complexes or municipalities, when it turn out to be convenient, so that the amount of the discharges needing an intensified treatment process can be reduced.

The degree of separation is dictated by technical economical considerations. Examples which may be given are the discharge of cooling water by industry and of rain water in municipalities which on many occasions would not be convenient to mix with other, more polluting discharges.

(5) Foreseeing the re-use of water when possible. This contributes to the reduction of the costs of pollution from two points of view.

In fact, even partial re-use of an industry's effluent would result in a smaller quantity of waste to be treated. At the same time, the demand for water would be smaller and therefore there would be a decrease in the load on the treatment plant. Similar advantages are produced by the re-use of intermediate effluents, such as the white waters in the paper industry and cooling water which, if not recycled, can be used for other purposes before being discharged. An example relating to the sugar industry is reported in Figure 1.

However' it must be admitted that in its present state the designing of treatment plants presents numerous difficulties, such as:

(1) Once it has been established which type of treated effluent is desired, the choice of a process scheme suitable for attaining the required results is not immediate, neither is it always certain.

(2) The dimension of the plant itself cannot be obtained on the basis of relations solely correlated with the characteristics of the effluent and influent and proved by experience.

(3) Once the final result desired has been established, precise analytical instruments adequate for defining it, are lacking. The case of the BOD can be taken as an example. It is perhaps the parameter which is most used for the definition of pollution and whose significance is very limited and, at times, quite misleading so that today more faith tends to be put in the TOC parameter.

3. Water Treatment Plant Design

As regards all chemical plants, even in the design of a water treatment plant, two clearly defined phases are distinguished: that of process and that of engineering.

As far as the first phase is concerned, in order to define it, it is best to begin with the old definition of A. D. Little:

Any chemical process, on whatever scale conducted may be resolved into a coordinate series of what may be called 'unit actions', the number of these basic unit operations is not very large and relatively few of them are involved in any particular process.

The process phase consists of the choice of unit operations, the establishing of the most suitable succession, and the definition of the exact conditions of operation, so as to obtain the desired result.

Once this first phase is finished' that of engineering takes over. This consists essentially of the dimensioning of the various equipment suitable for accomplishing the above-mentioned unit operations. In this phase, the work and the responsibilities of the chemical engineer are joined with those of the mechanical and civel engineer.

3.1. PROCESS ASPECTS

The list of unit operations increases with technical progress; thermal diffusion and electrodialysis are among the more recent acquisitions.

As regards the treatment of polluted water, it can be said that the terminology, the units of measurement, and the type of unit operations are often different from those of classical chemical engineering. For example in the latter, the term of unit operation, apart from the case of fermentation, is applied to processes in which the transformations are if not wholly, at least essentially physical. In the field of water treatment this limit has been exceeded and even operations in which chemical or biological transformations take place, are included in the field of unit operations.

In any case, from the conceptual point of view, nothing is changed. The introduction of operations of a biological type in the series of unit operations, depends only on the fact that some operations, for technical and economical reasons, are used essentially in the practice of water treatment and not in other fields of classical engineering.

The unit operations most frequently used in the field of treatment of polluted water are indicated briefly below and set out in Table I.

The operations in which gas transfer occurs are widely used in treatment plants especially in the secondary or tertiary phases of treatment. Gas transfer can take place, from the liquid phase to the gaseous phase (desorption) or vice versa (absorption) and is carried out by placing the two phases in contact in suitable conditions, across the widest surface possible. The gas phase can be at a pressure equal to, lower or greater than the atmospheric pressure, depending on the case.

There are very many examples and among them are the following: (see Table I).

– aeration or oxygenation for the maintenance or creation of aerobic conditions in which determinate micro-organisms may carry out their action;

– addition of air or oxygen for deferrization and demanganization;

– aeration for the removal of odorous or harmful substances.

The operations in which the transfer of solids takes place are clearly indicated in the table and have been written in decreasing order with respect to the dimensions of the solids which must be separated. It is useless to dwell on the forms and types of these operations which are widely used in nearly all chemical plants.

In the third category of the table a series of operations which is also very common in treatment plants, is grouped together. In these operations, even with some difference, a transfer of ions or molecules still takes place.

Under point (4) are grouped together all operations of an essentially biological nature, that is, those operations which take place with the more or less direct intervention of living organisms.

Depending on the operation to be carried out and therefore on the type of organism involved, it is necessary to create a suitable environment so that these organisms can carry out their functions without difficulty.

TABLE I

Different unit operations in waste water purification

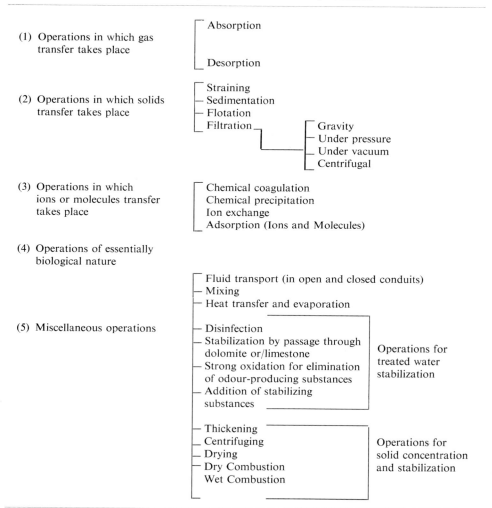

It is not simple to summarize briefly all the treatment operations in which living organisms intervene.

Perhaps it is best to refer to what happens in nature. Here essentially two types of organisms operate: the first type (autotrophic organisms) carry out their action by transforming simple substances, generally inorganic substances (ammonia, nitrites, nitrates, phosphates, silicates) into living organic substances and by-products. During this process light energy is generally utilized and gas is released (CO_2 and O_2).

Examples of this kind of organism are planctonic algae.

The second type (heterotropic organisms) convert complex substances, principally organic, into living cells and stable simple substances such as, for example, decomposition gas.

To this second type, other than bacteria and fungi, belong unicellular (protozoa) and pluricellular (Rotifera, nematoda, worms and insects) organisms. The bacteria and fungi are the first to attack the substances present in the waste water, beginning their decomposition. Thus they give origin to the food necessary for the linkage of the food chain. In practice, similar processes take place in the treatment plant although, almost always, under different conditions and times than with respect to the natural processes.

Those operations which are generally used for the stabilization of the plant's effluents have been emphasized so far. In fact it is necessary to point out that both the

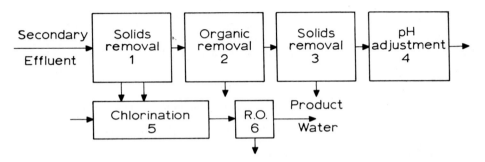

Fig. 2.   Reverse osmosis in a flow sheet for waste water purification.

treated water and the resulting sludge must be stabilized. The degree of stabilization depends on the final destination of the two streams.

The increasingly exacting demands on the characteristics of the treated water have recently been favouring the use of new techniques or ones which have already been tried out in other fields.

Here we are referring, in particular, to the processes which up till now have been used for the desalination of sea or brackish water and which are now becoming applicable in the field of water purification, especially in the tertiary phase of treatment. Among these, those which deserve particular attention are the processes of ion exchange and those in which membranes are used, i.e., electrodialysis and reverse osmosis.

As regards reverse osmosis, it seems that this type of process proves to be particularly suitable if applied to the treatment of waste water in that the substances to be removed are of greater dimensions than those in sea or brackish water and moreover the operation pressures are smaller. In the case of treatment however, there are some added difficulties essentially concerning fouling or plugging of the equipment and micro-biological contamination. These could be avoided by making the plant operate in suitable conditions indicated for example in the flow-sheet in Figure 2.

3.2. ENGINEERING ASPECTS

In the classical chemical industry the apparatus in which unit operations take place is designed with formulae and data drawn from long experience which are increasingly precise and reliable. Designing with more precise formulae, other conditions being equal, means obtaining, after all, a better product or one which costs less.

Unfortunately, as far as plants for the treatment of polluted water are concerned, the difficulties which face exact designing are many. For this reason, it has been said by some authors that pollution control is more an art than a science. If this is true, with all due respect for art, we hope that the situation is temporary.

The first condition for allowing the correct designing of a plant is that the physical and chemical characteristics of the various streams and the variation of the flow rate

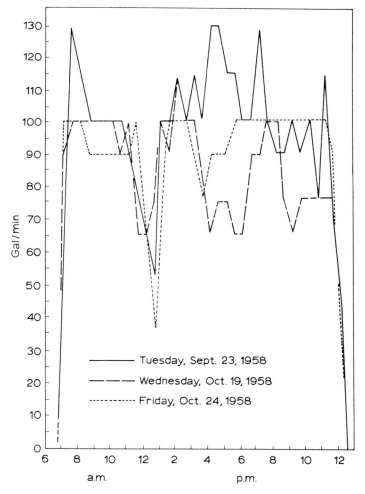

Fig. 3. Waste-water flow, weir No. 1 (Nemerov, 1963).

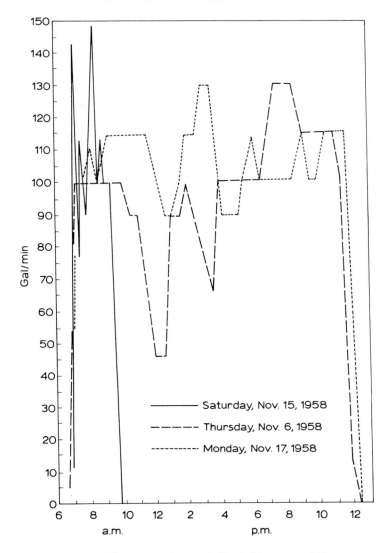

Fig. 4.   Waste-water flow, weir No. 2 (Nemerov, 1963).

should be definite. While it is not excessively difficult to establish, at least roughly, the characteristics of the effluents, considerable difficulties are encountered for the influents.

In fact, as regards the effluents, in practice it is only necessary to establish up to what point it is convenient to take the purification; either, for example, it is sufficient to obtain a partially purified effluent to be let into a water-course or it is best to obtain water which is directly re-useable from an urban, agricultural or industrial complex. These decisions will have to be taken on the basis of technical-economical considera-

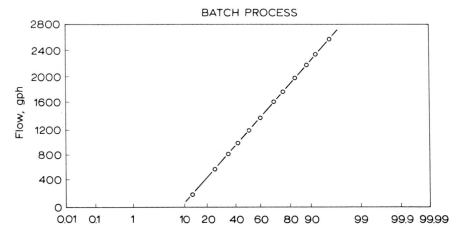

Fig. 5. Probability of occurrence of a certain flow rate in a typical batch process.
(According to Eckenfelder, 1966.)

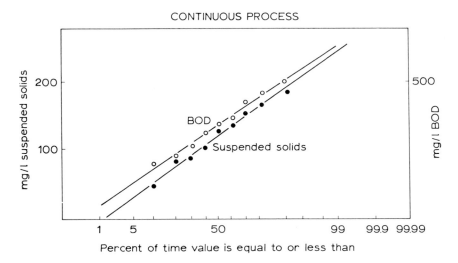

Fig. 6. Probability of occurrence of BOD and suspended solids in raw waste. (According
to Eckenfelder, 1966.)

tions which take account of the water and ecological situation of the zone being
examined. In particular, it is necessary to take the following into account:

– Government regulations in force in the zone under consideration, with particular
reference to quality standards.

– The condition of the water course which will receive the treated discharges. In
this stage an examination will have to be made of the capacity of the receiving water
to assimilate the discharge which is treated on the basis of the potential use for which
the water itself has been intended (bathing, fishing, irrigation, etc.). For example it

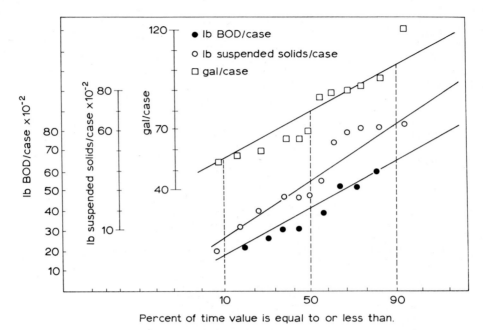

Fig. 7.   Daily variation in waste flow and characteristics: tomato waste. (According to
Eckenfelder, 1966.)

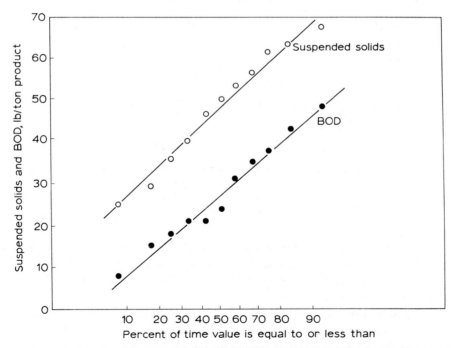

Fig. 8.   Variations in suspended solids and BOD from eleven paper board mills. (According
to Eckenfelder 1966.)

is possible for large water courses, intended for uses which are not particularly valuable, to receive lightly treated discharges without difficulty. This would be impossible for small water courses especially if they are subject to seasonal fluctuation.

– Possible uses for the effluent. If for example in the zone under examination there is a need for irrigation water, it may be convenient to carry out a more intensified treatment and to use the plant's effluent for this purpose.

On the basis of these and other considerations, the characteristics of the stream, i.e., the treated water, which will be discharged from the plant are established case by case.

However, depending on the variability of the operational conditions, the real characteristics may differ to a greater or lesser degree from those forecast. It is therefore necessary, especially for plants of considerable dimension, that the entity and duration of the allowable shifting from the forecasts should be determined in connection with the intended use of the effluent and with the characteristics of the receiving water body.

As regards the influents, unless it is a matter of a treatment plant for a single industry with simple production processes, it will be necessary to carry out a vast and, at times, complex preliminary analysis to establish the data on which the plant can be dimensioned with the entity and duration of the foreseeable variations

Eloquent examples of some situations are illustrated in the following diagrams and table. (See Figures 3–8; Table II and III.)

Figures 3 and 4 represent the flow variations during the day of the wastes coming from an industry which manufactures electrical and mechanical business machines.

TABLE II

Waste-water, influent at city disposal plant. (According to Nemerov, 1963.)

Month	1956		1957		1958		1957	1958
	Average flow, mgd	Suspended solids, ppm	Average flow mgd	Suspended solids, ppm	Average flow, mgd	Suspended solids, ppm	BOD, ppm	BOD, ppm
January	5.25	138	5.25	66	5.20	95	24	25
February	5.19	135	5.39	78	5.06	87	29	108
March	8.47	70	6.06	82	7.31	45	30	77
April	8.67	53	6.36	72	8.30	75	32	328
May	6.39	81	5.55	95	7.12	124	39	132
June	5.33	80	4.98	89	6.47	59	44	88
July	0.71	94	4.69	112	5.23	98	42	80
August	4.07	129	4.26	99	4.57	99	64	142
September	3.86	128	3.94	174	4.44	90	44	131
October	4.17	139	3.59	135			71	
November	3.84	123	3.36	138			61	
December	4.83	92	4.58	150			74	
Averages	5.40	105	4.86	108			46	

concentration gradient in apparent defiance of Fick's law. For this to be possible there must be some differentiation between the two sides of a membrane so that the binding of an ion to a carrier molecule is differently affected. In this connection Marek, while on leave from Prague at Waterloo (Canada), pointed out an extremely simple example of a diffusion and reaction problem whose solution is asymmetrical in spite of symmetrical boundary conditions. Such a reaction is capable of providing just this kind of differentiation between the two sides of the membrane and is being investigated further.

Years ago Turing suggested a problem in morphogenesis and raised the question of whether pattern and structure could arise out of diffusion and reaction in a surface. The surface has the advantage that it can be embedded in three dimensional space and hence apparent creation in the surface can arise by transport from the third dimension. This gives the normally staid reaction operator a certain wildness not open to it in three dimensions by introducing terms that apparently violate the condition of microscopic reversibility. My colleague Scriven and several of his students have shown how pattern can arise in this kind of system and recently with deSimone has actually produced chemically oscillating patterns in a thin layer. Prigogine and others have also been looking at open systems in which structure tends to arise spontaneously and Segal has produced a most interesting theoretical basis for chemotaxis.

Another area where a chemical engineer's interest in control and information science may lead him into biological contacts is that of systems theory. The aim of the systems theorist (according to R. E. Kalman, one of the great masters of the art and science of systems theory) is "to create systems which approach, or perhaps even surpass, capabilities normally observed only in the living world". The systems theorist may not provide ready made models for the brain, but he does hope that some of the methods that he may develop will be of use in understanding it, and if he is a good theorist, he should be able to provide a useful critique of the models that embody experimental results. In certain specialised areas, such as that of linear systems theory, results of great depth and penetration have been obtained. More recently the general algebraic basis of chemical kinetics has been investigated by Feinberg, by Jackson and Horn, and by Sellers. The general mass action kinetics discussed by Jackson and Horn provide a basis for discussing the possibility of oscillating reaction schemes such as those that Higgins has examined.

The whole area of enzyme technology has scarcely been mentioned, though it has recently received attention from the Engineering Chemistry Division of the National Science Foundation – particularly from the insight of L. G. Mayfield – and it now features in the latest thrust of research applied to national needs (RANN program). Major programs are under way in various centers including one under A. Humphrey at Pennsylvania. Mention should also be made of Professor Elmer Gaden of Columbia, one of the founders of research in fermentation engineering, and of Professor Lilly and his team in England. A natural point of contact is the common interest of the chemical engineer and biochemist in the kinetics of reactions and there is an excellent review of 'enzyme kinetics and engineering' in the opening pages of Volume 18 of the *A.I.Ch.E. Journal*.

Let me end this survey as I began it by reminding you that it is eclectic to a degree, at best suggestive, but, in view of the proportions to which the subject has already grown, inevitably sketchy. The following references are offered in the same spirit.

References

Tracers and Residence Time Distributions

Aris, R.: 1966, in K. B. Warren (ed.), *Intracellular Transport*, Academic Press, New York.
Danckwerts, P. V.: 1953, *Chem. Eng. Sci.* **2**, 1.
Klinkenberg, A.: 1964, *Trans Inst. Chem. Eng. London* **43**, 141.
Levenspiel, O. and Bischoff, K. B.: 1963, *Advan. Chem. Eng.* **4**, 95, Academic Press, New York.
Sheppard, C. W.: 1962, *Basic Principles of the Tracer Method*, Wiley, New York.
Spalding, D. B.: 1958, *Chem. Eng. Sci.* **9**, 74.

Taylor Diffusion

Aris, R.: 1956, *Proc. Roy. Soc. London Ser. A* **235**, 67.
Chatwyn, D. C.: 1970, *J. Fluid Mech.* **43**, 321.
Gill, W. N.: 1967, *Proc. Roy. Soc. London Ser. A* **298**, 335.
Gill, W. N. and Sankarsubramanian, R.: 1970, *Proc. Roy. Soc. London Ser. A* **316**, 341 and 1971, **322**, 101.
Taylor, G. I.: 1953, *Proc. Roy. Soc. London Ser. A* **219**, 186.
Taylor, G. I.: 1953, *Appl. Mech. Rev.* **6**, 256.

Drug Distribution

Bischoff, K. B. and Dedrick, R. L.: 1969, *Chem. Eng. Prog. Symp. (Ser. No. 84)* **64**, 32.
Bischoff, K. B. and Dedrick, R. L.: 1970, *J. Theor. Biol.* **29**, 63.
Bischoff, K. B., Dedrick, R. L., and Zaharko, D. S.: 1970, *Proc. Ann. Conf. Med. Biol.* **12**, 89.
Dedrick, R. L., Gabelnick, H. L., and Bischoff, K. B.: 1968, *Proc. Ann. Conf. Med. Biol.* **10**, 36.1.

Population Studies

Fredrickson, A. G.: 1971, *Math. Biosci.* **10**, 117.
Fredrickson, A. G., Ramkrishna, D., and Tsuchiya, H. M.: *Math. Biosci.* **1**, 327.
Ramkrishna, D., Fredrickson, A. G., and Tsuchiya, H. M.: 1966, *J. Gen. Appl. Microbiol.* **12**, 311.
Tsuchiya, H. M., Fredrickson, A. G., and Aris, R.: 1966, in *Advan. Chem. Eng.* **6**, 125.

Chromatography

Bayle, G. G. and Klinkenberg, A.: 1954, *Rec. Trav. Chim. Pays-Bas Belg.* **73**, 1037.
deVault, D.: 1943, *J. Am. Chem. Soc.* **65**, 532.
Glueckauf, E.: 1946, *Proc. Roy. Soc. London* **A186**, 35.
Rhee, H. K., Aris, R., Amundson, N. R.: 1970, *Phil. Trans. Roy. Soc. London* **A267**, 419, and 1971, **A269**, 187.
Sillén, L. G.: 1950, *Ark. Kemi. Miner. Geol.* **2**, 477.

Transport Phenomena

Keller, K. H. and Friedlander, S. K.: 1966, *J. Gen. Physiol.* **49**, 663.
Mitchell, P.: 1970, in H. P. Charles and J. G. Knight (eds.), *Organisation and Control in Prokaryotic and Enkaryotic Cells*, Cambridge Univ. Press.
Whittam, R.: 1964, *Transport and Diffusion in Red Blood Cells*, Williams and Wilkins, Baltimore.

Pattern Formation and Chemotaxis

Edelstein, B. B.: 1970, *J. Theor. Biol.* **26**, 227.
Keller, E. F. and Segal, L. A.: 1970, *J. Theor. Biol.* **26**, 339, and 1971, **30**, 225 and 235.
Scriven, L. E. and Gmitro, J. J.: 1966, in K. B. Warren (ed.), *Intracellular Transport*, Academic Press, New York.
Scriven, L. E. and Othmer, H.: 1969, *Ind. Eng. Chem. Fundam.* **8**, 303.
Turing, A. M.: 1952, *Phil. Trans. Roy. Soc. London Ser.* B **237**, 37.

Insolubilized Enzymes

Kostin, M. D. and Carbonell, R. G.: 1972, *A.I.Ch.E. Journ.* **18**, 1.
Mosbach, K. and Mattiason, B.: 1970, *Acta Chem. Scand.* **24**, 2093.
Shuler, M. L., Aris, R., and Tsuchiya, H. M.: 1972, *J. Theor. Biol.* **35**, 67.
Silman, I. H. and Katchalsky, E.: 1966, *Ann. Rev. Biochem.* **35**, 873.

Systems Theory and General Topics

Bartholomay, A. F.: 1968, *Bio Sci.* **18**, 717.
Bunge, M.: 1968, *Models in Theoretical Science*, Akten des XIV Int. Kong. f. Philosophie, Wien.
Bunge, M.: 1969, 'Analogy, Simulation, Representation', *Rev. Int. de Philosophie (Brussels)*, 23e year, No. 87.
Feinberg, M.: 1972, 'On Chemical Kinetics of a Certain Class', *Arch. Rat. Mech. Anal.* **46**, 1.
Higgins, J.: 1967, *Ind. Eng. Chem.* **59**, 18.
Horn, F. and Jackson, R.: 1972, 'General Mass Action Kinetics', *Arch. Rat. Mech. Anal.* **47**, 81.
Kalman, R. E.: 1968, in M. D. Mesarovic (ed.), *Systems Theory and Biology*, Springer-Verlag, New York.
Kalman, R. E., Falb, P. L., and Arbib, M. A.: 1969, *Topics in Mathematical Systems Theory*, McGraw Hill, New York.
Nooney, G. C.: 1965, *J. Theoret. Biol.* **9**, 239.
Sellers, P.: 1967, *SIAM J. Appl. Math.* **15**, 13.

Appendix

Diffusion and Electrostatic Hindrance with Insolubilized Enzymes

The current popularity of insolubilized enzymes opens up immediately an area to which the chemical engineer can contribute something, for he has thirty years of experience in the problems of diffusion and reaction. Here with the enzyme problem he is confronted with an emphasis on a new class of kinetics and I would like to take one fairly straightforward problem in this area, the first part of which is due to Shuler (see reference under *Insolubilized Enzymes*).

An enzyme E is attached to the surface of a spherical bead in a charged medium of ionic strength I. Far from the bead the concentration of substrate is uniform and equal to s_f, while at the surface it is s_0 and forms a product at a rate

$$v = V_m s_0/(s_0 + K_m).$$

K_m is the Michaelis constant and V_m the maximum rate per unit area.

The flux of substrate to the surface is

$$J_s = D \frac{ds}{dr} + s \frac{zDF}{RT} \frac{d\psi}{dr}$$

where $\psi(r)$=electrostatic potential, F=Faraday's constant, D=diffusivity of substrate, T=temperature, R=gas constant, z=valency of substrate.

It is convenient to write

$$\psi_0 = \psi(0),$$
$$\lambda = zF\psi(0)/RT.$$

so that

$$J_s = D \left\{ \frac{ds}{dr} + \lambda s \frac{1}{\psi_0} \frac{d\psi}{dr} \right\}.$$

Now in the space around the bead

$$\frac{1}{r^2} \frac{d}{dr} (r^2 J_s) = 0.$$

At the surface $r=a$ of the bead

$$J_s = V_m s_0/(s_0 + K_m) = R \quad \text{(say)}.$$

Thus

$$D \left(\frac{ds}{dr} + \lambda s \frac{1}{\psi_0} \frac{d\psi}{dr} \right) = R \frac{a^2}{r^2}.$$

Multiplying by $\exp[\lambda\psi(r)/\psi_0]/D$ gives

$$\left[\frac{d}{dr} \int s e^{\lambda\psi(r)/\psi_0} \right] = \frac{R}{D} \frac{a^2}{r^2} e^{\lambda\psi(r)/\psi_0},$$

and integrating from $r=a$ to $r=\infty$, where $\psi(r)=0$

$$s_f - s_0 e^{\lambda} = \frac{Ra}{D} \int_0^{\infty} e^{\lambda\psi(a\varrho)/\psi_0} \frac{d\varrho}{\varrho^2}.$$

Thus if

$$v = K_m/s_f.$$

$$\mu = \frac{V_m a}{Ds_f} \int_1^{\infty} e^{\lambda\psi(a\varrho)/\psi_0} \frac{d\varrho}{\varrho^2},$$

then

$$1 - \sigma e^{\lambda} = \frac{\mu\sigma}{v + \sigma},$$

where $\sigma = s_0/s_f$. This would give us an equation for σ (i.e. s_0) from which the reaction rate could be calculated, but it is easier to write one for the reaction rate itself. Indeed, let us introduce a typically chemical engineering concept, the effectiveness factor. If s_f were the concentration at the surface, the reaction rate would be

$$V_m 4\pi a^2 s_f/(s_f + K_m) = V/(1 + v).$$

In fact it will be $V_m 4\pi a^2 s_0/(s_0 + K_m) = V\sigma/(\sigma + v)$. Let the effectiveness factor be the ratio of these

$$\eta = \frac{\sigma(1 + v)}{\sigma + v}.$$

By the previous equation

$$\eta = \frac{1 + v}{\mu}(1 - \sigma e^\lambda),$$

and eliminating σ we obtain

$$\mu\eta^2 - \eta(1 + \mu + ve^\lambda)(1 + v) + (1 + v)^2 = 0.$$

This quadratic can easily be solved for η but when μ is small it is given by

$$\eta = \frac{1 + v}{1 + ve^\lambda} - \mu\frac{v(1 + v)e^\lambda}{(1 + ve^\lambda)^3} + 0(\mu^2).$$

To the same degree of approximation this is

$$\eta = \frac{1 + v}{1 + ve^\lambda[1 + \mu/(1 + ve^\lambda)]}.$$

So that the rate of reaction is

$$\frac{\eta v}{1 + v} = \frac{Vs_f}{s_f + K'_m}$$

where

$$K'_m = K_m e^\lambda\left\{1 + \frac{V_m a}{D(s_f + K_m e^\lambda)}\int_1^\infty e^{\lambda\psi(a\varrho)/\psi_0}\frac{d\varrho}{\varrho^2}\right\}.$$

Thus for small μ the effect of diffusion is taken care of by modifying the Michaelis constant. This was proposed by Hornby, Lilly, and Crook but their modifying factor is incorrectly computed. The true relationship between η and the parameters λ, μ, v may be presented by writing the quadratic in the form:

$$\mu = \frac{1}{\eta/(1 + v)} - \frac{ve^\lambda}{1 - \eta/(1 + v)},$$

and is shown as contours of $\eta/(1 + v)$ in the plane of μ and ve^λ (Figure 1). The value

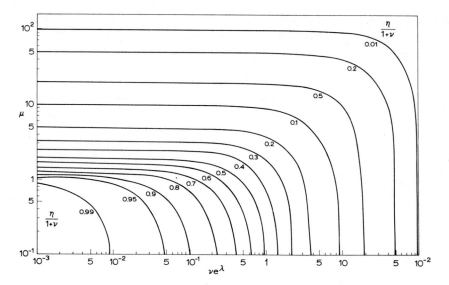

Fig. 1.   Contours of $\eta/(1+\nu)$ in the plane of μ and νe^{λ}.

of μ depends of course upon the expression chosen to represent the electrostatic field around the bead.

It is interesting to note a two dimensional surface with s constant at large distances from a circular patch is not sufficiently vast to sustain a continual reaction at the boundary of a small circle. For consider r to be the distance from the centre of a circular patch of radius a in the plane and ignore the electrostatic effects. Then

$$\hat{J}_s = \hat{D} \frac{d}{dr},$$

where $\hat{\ }$ denotes quantities defined in the surface. But in two dimensions

$$\frac{d}{dr}(r\hat{J}_s) = 0,$$

so that

$$\hat{J}_s = \frac{a}{r}\hat{R} = \frac{1}{r}\frac{\hat{V}_m \hat{s}_0}{\hat{s}_0 + \hat{K}_m}.$$

Now if

$$\frac{d\hat{s}}{dr} = \frac{a}{r}\frac{\hat{R}}{\hat{D}},$$

we see that $\hat{s} \to \infty$ as $r \to \infty$ and the boundary condition $\hat{s} \to \hat{s}_f$ could not be met.

Instead if we are to sustain such a reaction the substrate must be adsorbed on the surface from a third dimension. Let $\hat{S}$ be a total coverage and suppose that the net rate of adsorption per unit area is

$$k_a s_f (\hat{S} - \hat{s}) - k_d \hat{s}.$$

Then

$$\frac{\hat{D}}{r} \frac{d}{dr}\left(r \frac{d\hat{s}}{dr}\right) - (k_a s_f + k_d)\,\hat{s} = - k_a s_f \hat{S},$$

and we can let

$$\hat{s} \to \hat{s}_e = \hat{S} \frac{k_a s_f}{k_a s_f + k_d} = \hat{S} \frac{\Sigma}{1 + \Sigma},$$

where

$$\Sigma = k_a s_f / k_d.$$

Let

$$\alpha = a\,\{(k_d + k_a s_f)/\hat{D}\}^{1/2},$$

and

$$u = \hat{s}/\hat{S}, \qquad \varrho = r/a,$$

then

$$\frac{1}{\varrho} \frac{d}{d\varrho}\left(\varrho \frac{du}{d\varrho}\right) - \alpha^2 u = - \alpha^2 \Sigma/(1 + \Sigma).$$

The solution of this equation is

$$u = \frac{\Sigma}{1 + \Sigma} - \left(\frac{\Sigma}{1 + \Sigma} - u_0\right) \frac{K_0(\alpha\varrho)}{K_0(\alpha)},$$

where K_0 is the second kind of modified Bessel function of order zero. The first kind cannot be present as it tends to infinity as its argument tends to infinity.

But at the edge of the patch

$$\hat{D} \frac{d\hat{s}}{dr} = J_s = \frac{\hat{V}_m \hat{s}_0}{\hat{K}_m + \hat{s}_0},$$

so that at $\varrho = 1$

$$\frac{du}{d\varrho} = \frac{a \hat{V}_m}{\hat{D}} \frac{u_0}{\hat{\kappa} + u_0} = \hat{\mu} \frac{u_0}{\hat{\kappa} + u_0}.$$

But

$$\frac{du}{d\varrho} = \alpha \left(\frac{\Sigma}{1 + \Sigma} - u_0\right) \frac{K_1(\alpha)}{K_0(\alpha)}.$$

Let

$$e^\lambda = \frac{1 + \Sigma}{\Sigma}, \qquad \mu = \hat{\mu} \frac{1 + \Sigma}{\Sigma} \frac{K_0(\alpha)}{\alpha K_1(\alpha)}, \qquad v = \hat{\kappa},$$

and we have the same quadratic as before.

FLOCCULATING AGENTS IN WASTE WATER PURIFICATION AND SLUDGE DEWATERING

STIG FRIBERG and KELVIN ROBERTS

The Swedish Institute of Surface Chemistry, Stockholm, Sweden

Abstract. It has been pointed out that many of the problems encountered in water purification are of colloid and surface chemical character. In fact this field exemplifies many of these questions in surface and colloid chemistry, satisfactory answers to which have not yet been found.

Among these are the mutual influence of different polymers on precipitation phenomena, manifested in the influence of the pollution level of waste water on the sedimentation efficiency of aluminium hydroxide phosphate, and the even more complex inter-relationships playing a role when the dewatering of sludge takes place.

Waste water purification has since the earliest times utilized accelerated biological degradation of the organic material present by using activated sludge. Microorganisms which developed in sludge were re-used and the induction time for obtaining rapid bacterial growth was reduced. The purification efficiency was rather low, removing about half of the organic impurities. As a rule the recipients could fairly easily sustain the charge, and water quality was high. The increased output of phosphorus and nitrogen compounds from urban areas has, however, exceeded the capacity of the natural cleaning processes in the recipient. This has caused a rapid growth of green algae, which in turn reduced oxygen photosynthesis, causing severe viability problems.

Removal of phosphorus by chemical precipitation using aluminium sulphate has sufficed to prevent rapid algal growth, the high nitrogen load obviously having to be combined with an adequate supply of phosphorus in order to be nutritionally efficient.

In principle, the chemical reaction at low pH:s for the precipitation process is simple, the aquoaluminium complex is deprotonized to the neutral insoluble aluminium hydroxide. Aluminium phosphate is co-precipitated giving an overall reaction:

$$Al(H_2O)_i^{3+} + PO_4^{3-} \xrightarrow{\text{inc. pH}} Al_x(OH)_y(PO_4)_z + 1(H_2O),$$

where $x = y/3 + z$.

Sometimes the precipitation reaction does not take place, or the precipitate formed is loose and slimy, containing large amounts of water. This precipitate, which also contains organic impurities, forms the sludge of the purification process and shows a water content of between 95–99% (Bacon and Dalton, 1966). The water in the sludge is extremely difficult to remove; Figure 1 shows the extremely high temperatures which have to be used. The water is obviously strongly bound to various compounds in the sludge, some of the interparticulate water occurs in the form of slimy gels (Figure 2). In these, water is helf by polymers or by associated structures of amphiphilic compounds (Figure 3). One example of such an association phase (Friberg *et al.*,

G. Lindner and K. Nyberg (eds.), Environmental Engineering, 227–236. All Rights Reserved

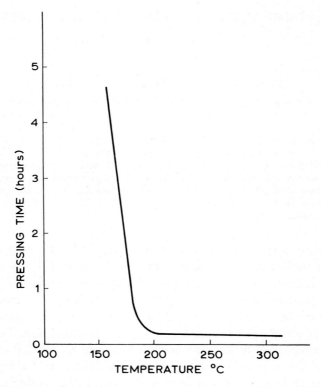

Fig. 1.   Temperature effect on required pressing time after heat treatment of sludge (according to Malina, 1971).

1969) containing 85% by weight of water was ultra-centrifuged with a field of 300000 g for 48 hours without any water separation.

With regard to the general importance of this problem and to the fact that so little is generally known about the basic factors which determine the strong 'bonding' of water in sludges, we here present a basic description of the underlying phenomena, followed by examples of how knowledge of these might be utilized in practice. Water is in fact not especially strongly bound to any particles in sludge. It is rather the effect of the particles or the aggregates being prevented from approaching each other; the water is present in the interparticulate space and 'bound' mainly by its own high intermolecular forces manifested in its high surface energy. The prevention and facilitation of particle approach are treated in the next section.

1. Stable Suspensions of Particles in Water

In general all particles in water are attracted to each other due to interaction between the alternating fields created by the rapid change of electron distribution in the atomic and molecular orbitals. For small spherical particles close to each other in a vacuum

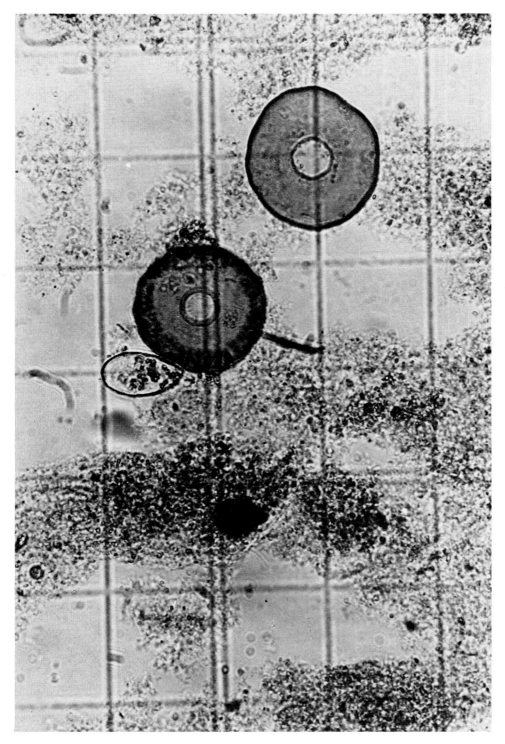

Fig. 2.   Microphotograph of sludge from Käppala purification plant, Stockholm.

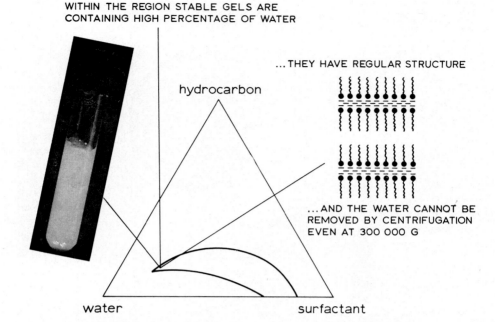

WITHIN THE REGION STABLE GELS ARE
CONTAINING HIGH PERCENTAGE OF WATER

...THEY HAVE REGULAR STRUCTURE

hydrocarbon

...AND THE WATER CANNOT BE
REMOVED BY CENTRIFUGATION
EVEN AT 300 000 G

water   surfactant

Fig. 3.   Water is strongly bound in gels of biopolymers or in liquid crystalline phases containing
lipids and surfactants.

the potential depends on the distance (Kruyt, 1952)

$$V_{attr.} = -\frac{A}{12}\frac{a}{d},$$

where A is a constant, $\sim 10^{-13}$ erg, a is the particle radius and d is the distance between the particle surfaces. The attraction potential is low; the force between two
spheres of radius 10 cm and 1 mm apart will only be $\sim 10^{-11}$ dyn, but, anyhow, sufficient to cause a rapid flocculation, coagulation, and sedimentation of small particles
in water if they are not protected.

The most general protection mechanism is an electrical charge on the surfaces
giving rise to a repulsion potential when two particles approach each other. The
repulsion potential is a complicated function of charge density and distance; a crude
approximation gives

$$V_{rep.} = K\,\xi^2 e^{-Hd}$$

where ξ is the electric surface potential, H is approximately proportional to the charge
of the electrolyte ions with opposite charge to the particle surface times the square
root of their concentration and d is the distance. Figure 4 demonstrates that the

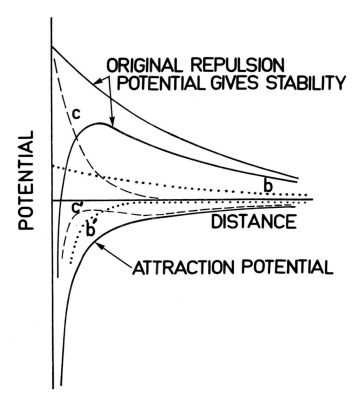

Fig. 4. Reduction of the surface potential (b) or compression of the electric double layer (c) both reduce the electric repulsion potential and cause destabilization of the system; c′ and b′ represent total potential, c and b represent repulsion potential.

repulsion potential might be reduced by reduction of the surface potential or by an increase in the electrolyte charge or concentration.

2. Coagulation and Flocculation

In practice this means that the particles might be destabilized by the addition of salt (c) or by surface active substances with opposite charge to the surface (b) (Figure 4). Hydrolyzed metal ions also adsorb strongly to the surface changing the surface potential (Matijević and Force, 1970).

Another extremely efficient way of flocculating particles consists of the addition of polymers with charged groups. The mechanism (Clayfield and Lumb, 1966; La Mer and Healy, 1966) is considered to be a molecular bridging between the particles by loops of the polymer chain. Figure 5 demonstrates the mechanism. It is worth noting that a complete coverage of the particle by the polymer with loops protruding into the solution, effects an efficient *protection* against flocculation.

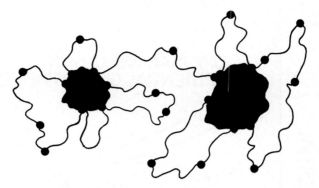

(a) INCOMPLETELY COVERED SURFACE
GIVES BRIDGING AND FLOCCULATION

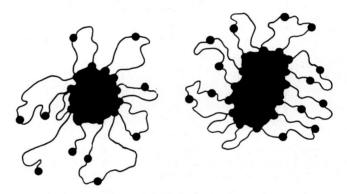

(b) COVERED SURFACES GIVE
PROTECTION

Fig. 5.   Schematic representation of polymer adsorption leading to (a) bridging and flocculation, (b) protection.

3. Application Problems

3.1. EFFICIENCY OF SEDIMENTATION

The removal of phosphorus is critically dependent on the flocculation and sedimentation of aluminium hydroxide phosphate. In practice it turned out that the sedimentation and purification efficiency varies with time within a given plant and that plant results differ considerably from those predicted from laboratory tests.

In our opinion this might be explained by the fact that the pollution level of waste water has a pronounced influence on the flocculation and sedimentation behaviour (Friberg and Roberts, 1972).

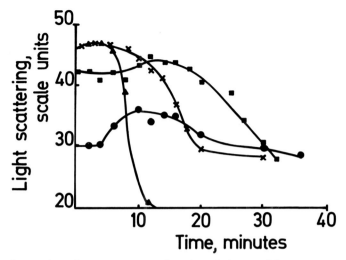

Fig. 6. The sedimentation of wastewater as a function of time at different wastewater dilutions. 0.4 g l^{-1} aluminium sulphate is added at time $t = 0$, and 10 ml l^{-1} Purifloc A23, an anionic poly-acrylamide, at time $t = 4$ min. ▲ – No dilution, × – Diluted twice, ■ – Diluted four times, ● – Diluted eight times.

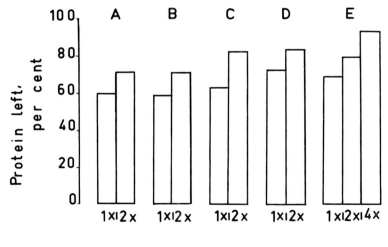

Fig. 7. The percentage removal of proteins from wastewater as a function of wastewater dilution, using 0.4 g l^{-1} aluminium sulphate. A – no flocculant, B – 1 mg l^{-1} Sedipur KA, C – 10 mg l^{-1} Sedipur KA, D – 1 mg l^{-1} Purifloc A23, E – 10 mg l^{-1} Purifloc A23.

The waste water sedimentation rate using a fixed dosage of aluminium sulphate and polyacrylamide varies when the concentration of dissolved impurity material in the waste water varies. This is demonstratee by Figure 6, where twice dilution of the waste water results in an increase of the retention time between polymer addition and onset of sedimentation from 1 to 4 min, and also reduces the rate of sedimentation as shown by the reduced (negative) slope of the turbidity vs time curve. Dilution by 4 and 8 times, respectively, shows continuation of this trend.

Percentage removal of dissolved proteins also depends on the waste water dilution as shown in Figure 7. Similar data have been obtained for dissolved saccharides and lipids, showing that each of these dissolved impurities is involved in the chemical sedimentation process.

Investigations in a model system using E-coli bacteria have shown that proteins increase, whilst saccharide and lipid decrease, the sedimentation rate observed when aluminium hydroxide and polyacrylamides are used for chemical sedimentation. Since the concentrations of each of these respective organic materials in domestic waste water varies periodically by as much as a factor of 5, it is clear that coagulant

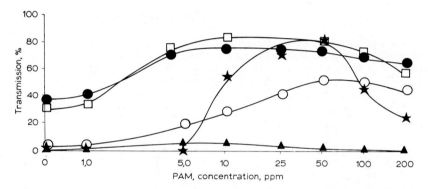

Fig. 8.   The sedimentation of 10 g l⁻¹ kaolin suspensions with anionic polyacrylamide at pH 5 in the presence of different amounts of pre-added aluminium ion.
▲ – 0 ppm aluminium ions, ● – 5 ppm aluminium ions, □ – 10 ppm aluminium ions, ○ – 20 ppm aluminium ions, ★ – 30 ppm aluminium ions.

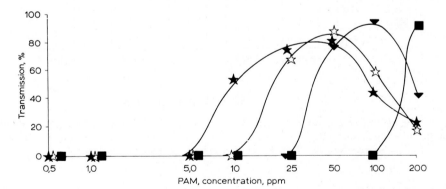

Fig. 9.   The sedimentation of 10 g l⁻¹ kaolin suspensions with anionic polyacrylamide at pH 5 in the presence of different amounts of pre-added aluminium ion.
★ – 30 ppm aluminium ions, ☆ – 40 ppm aluminium ions, ▼ – 50 ppm aluminium ions, ■ – 100 ppm aluminium ions.

and flocculant dosage should be correspondingly adjusted to obtain continual optimal sedimentation in practice.

3.2. HYDROLYSED METAL IONS AND POLYACRYLAMIDES

At pH values between 4 and 5, aluminium salts form soluble hydrolysed species of probable type $Al_8(OH)_{20}^{4+}$ (Matijević 1970). In addition to the effects of adsorption and charge reversal at the surfaces of negatively-charged suspended particles already described, these hydrolysed species have a specific and decisive influence on sedimentation when anionic polyacrylamides are employed. The sedimentation of kaolin with aluminium ions and anionic polyacrylamide at pH 5 is shown in Figures 8 and 9. (A pH at 5 is widely used for sizing in the paper industry, and therefore occurs often in white water systems in paper mills.) In the absence of aluminium, sedimentation is poor, while small amounts, 5 and 10 ppm aluminium, give good sedimentation at polyacrylamide concentrations from 0.1 to 100 ppm. Larger amounts of aluminium, 30 ppm or above, result in poor sedimentation till 50 ppm or more polyacrylamide has been added.

The clear practical conclusion (Roberts *et al.*, 1972) is that aluminium dosage must be kept below 20 ppm to allow efficient sedimentation at low polymer dosage and low cost. The lack of sedimentation of kaolin in the absence of aluminium can be explained by weak adsorption of the negatively-charged polymer at the surface of the negatively-charged play.

Hydrolysed aluminium ions bind strongly to the anionic polymer, leading to sedimentation of the polymer itself at sufficient concentrations. The production of an aluminium polymer sediment is responsible for the sedimentation maxima observed at aluminium dosages greater than 30 ppm and polymer dosages greater than 40 ppm.

The high efficiency of sedimentation when 5 or 10 ppm polymer aluminium is added at low polymer dosage is thought to be due to the improved polymer adsorption via bonding to the dydroxy-aluminium complexes adsorbed at the clay surfaces.

3.3. SLUDGE DEWATERING PROBLEMS

The sludge contains between 95 and 99.5% water, strongly held in the structure (Figure 1). In future, sludge problems will become even more important. From the economic point of view the problem is serious enough today; of the total cost for a waste water purification plant, between 21–65% is necessary for the sludge treatment (Randall *et al.*, 1971).

Of all the methods which have been used to remove the water viz. heat treatment (Malina, 1971), addition of salts and other inorganics, freezing techniques (Gale and Baskerville, 1970) and addition of polyelectrolytes combined with filter pressing and centrifuging, only the last mentioned appear to have met with a reasonable level of success.

The immediate question is of course the structure of the most efficient polymer to be used for sludge dewatering. Although the theoretical treatment of the function of polyelectrolytes has been treated in a large number of articles (Clayfield and Lumb,

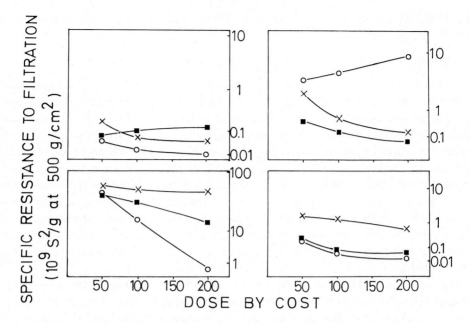

Fig. 10.   Dewatering of different sludges with three different polyelectrolytes.

1966; La Mer and Healey, 1966) and also several empirical investigations on the same theme exist, the present situation is that no simple rule exists for choice of a suitable polymer to dewater sludge.

The results from one investigation (Chen *et al.*, 1970) in which sixteen different polymers were used on four different sludges showed no systematic trend at all. Figure 10 shows that one polymer appeared superior for one sludge turned out to be the least efficient for another sludge.

References

Bacon, V. W. and Dalton, F. E.: 1966, *Pub. Works* **97**, 66.
Cheng, Ch., Updegraff, D. M., and Ross, L. W.: 1970, *Environ. Sci. Technol.* **4**, 1145.
Clayfield, E. and Lumb, E. C.: 1966, *J. Colloid Interface Sci.* **22**, 269.
Friberg, S. and Roberts, K.: 1972, *Ambio* (in press).
Friberg, S., Mandell, L., and Fontell, K.: 1966, *Acta Chem. Scand.* **23**, 1055.
Gale, R. S. and Baskerville, R. C.: 1970, *Filtr. Separ.*, No. 1, 37.
Kruyt, H. C. (ed.): 1952, *Colloid Sci.* **1**.
La Mer, V. K. and Healey, T. W.: 1966, *Rev. Pure Appl. Chem.* **13**, 112.
Malina, Jr., F.: 1971, *Water Res. Symp.*, Texas.
Matijević, E. and Force, C. C.: 1970, *Kolloid-Z. Z. Polym.* **225**, 33.
Overbeek, J. Th. G. and Verwey, S. V.: 1948, *The Stability of Lyophobic Colloids*, Elsevier, Amsterdam.
Randall, C. W., Turpin, J. K., and King, P. H.: 1971, *J. Water Pollut. Contr. Fed.* **43**, 102.
Roberts, K., Kowalewska, N., and Friberg, S.: 1972, *Proc. VIth Int. Conf. Surf. Act. Subst.* (in press).

PART VI

WATER POLLUTION

Industrial Examples and Engineering

INDUSTRIAL WASTE WATER TREATMENT BY PRECIPITATION AND ION EXCHANGE

SVEN ERIK JØRGENSEN

University of Pharmaceutists in Denmark, Copenhagen, Denmark

Abstract. Precipitation and ion exchange – especially in combination – are unit operations which are very promising for the future treatment of waste water with possibilities of reaching a high extent of purification, including substantial reduction of organic matter.

The mechanical-biological waste water treatment is the most important method which is used to meet a wide spectrum of waste water problems, including treatment of municipal waste water and several types of industrial waste water.

However, the mechanical-biological treatment is far from solving many rather essential waste water problems [17].

Increasing pollution due to urban growth and increased industrial activity [37–39] has stimulated the search for new methods to treat waste water.

Industry requires new processes to deal with waste water of special compositions, where conventional methods cannot be used. Space is expensive in industrial areas, therefore industry is looking for methods requiring less space. This has promoted the development of new compact plants [9, 34], as the usual lay-out of a biological plant requires large areas.

New demands on the purified water emphasize the need for new waste water treatment processes. This can be summarized in the following 4 points:

(1) to remove nutrients in order to avoid hypereutrofication of fresh waters [39],

(2) to remove toxic material, e.g., heavy metals, without any damage to the efficiency of the process, since the efficiency of the biological treatment suffers from the presence of this material [5, 14, 35],

(3) to remove microorganisms mainly due to the increasing use of surface water for production of drinking water [10]. The role of microorganisms in water used for recreative purposes have not yet been fully clarified [24, 25].

(4) to reduce the concentration of organic matter (measured as BOD_5, $KMnO_4$-number, COD or total carbon), since it is desirable to have the lowest possible concentration of organic matter in surface water used for the production of drinking water and because the organic matter also seems to influence hypereutrofication [3, 10, 15].

1. Chemical Precipitation as a Waste Water Treatment Process

Chemical precipitation is a unit operation used in technical scale for the treatment of waste water.

It is used to remove phosphates from waste water. The efficiency of the process is dependent on many variables, but should normally be between 75% and 95% [41].

G. Lindner and K. Nyberg (eds.), Environmental Engineering, 239–246. All Rights Reserved

The efficiency of this treatment on the receiving water is dependent on the factor which limits the growth of algae. [15, 32]. *Municipal Waste Water* is treated by precipitation with aluminium-sulphate. [2, 4, 12, 36, 42]

Precipitation can take place either immediately after the sand trap in the biological plant or after a mechanical-biological treatment. Precipitation is called respectively direct, simultaneous, or final [1, 11, 27, 31, 42].

Apart from obtaining a substantial reduction of phosphate concentration organic matter is also precipitated due to a reduction of the zetapotential [1]. Normally direct precipitation will reduce the $KMnO_4$-number and BOD_5 by 50–65% which has to be compared with the effect obtained by plain settling without the addition of chemicals – 25–40% [4].

TABLE I

Use of chemical precipitation for treatment of industrial waste water

Type of waste water	Chemical used	Ref. No.
Metal Plating and Finishing Industry	Lime	[35]
Iron Industry and Mining	Lime Aluminium sulphate	[16]
Electrolytical Industry	Hydrogen sulfide	[43]
Coke and Tar Industry	Lime or Sodium hydroxide	[44]
Cadmium Mining	Xanthates	[8]
Manufacturing of Glass- and Stone wool	Sodium hydroxide	[34]
Oil Refineries	Aluminium sulphate Iron (III) chloride	[45]
Manufacture of Organic Chemicals	Aluminium sulphate Iron (III) chloride	[46]
Photochemicals	Aluminium sulphate	[47]
Dye Industry	Iron (II) salts, Aluminium sulphate, Lime	[48]
Fertilizer Industry	PO_4^{3-}: Iron (II) salts, Aluminium sulphate, Lime NH_4^+: magnesium sulphate + phosphate	[49]
Plastics Industry	Lime	[50]
Food Industry	Ligninsulphonic acid, Dodecylbenzensulphonic acid, Glucose trisulphate, Iron (III) chloride, Aluminium sulphate	[19]
Paper Industry	Bentonite, Caoline, Starch, Polyacrylamid	[51]
Textile Industry	Bentonite, Aluminium sulphate	[20]

Recently chemical treatment has been used on a large scale in Sweden, the U.S.A., and other countries. However, it is an old process which was in operation in Paris in 1740 and in England in the middle of the 19th century. The so-called Langhlin process was used in the early thirties in the U.S.A. and just before the Second World War; Guggenheim introduced the process as a treatment preceding an activated sludge plant [31]. The method nevertheless did not find favour until the late sixties due to the cost of chemicals, low reduction in BOD_5, and increased sludge disposal problem.

The new claim mentioned in point 1, p. 239, has put new life into the process. The mechanical-biological treatment generally reduces the nutrient concentration by only 20–30% [13].

The production of biomass due to nutrients creates so-called secondary pollution which requires oxygen for oxidation of the organic matter as well as the organic matter in the waste water.

If it is presupposed that all phosphorous will promote the growth of algae, the oxygen demand due to the secondary pollution will be substantially higher than the BOD_5 [18, 29].

It should therefore be understood that it has been decided to treat waste water discharged into rivers and lakes chemically, although the problems of secondary pollution are not always that simple, that the removal of phosphates does not completely solve them [15].

Many types of industrial waste water can be treated by chemical precipitation with advantage.

Table I gives a survey of the use of this process on different types of industrial waste water.

The process has found wide application, particularly to remove metals from waste water from the metal industry and to reduce the amount of fiber in waste water from the paper industry.

2. Ion Exchange

The application of ion exchange has been greatly extended in the last few decades. Firms producing ion exchange resins have increased their production substantially. New types of resins are solving new problems or are more specifically suited to the problems to be solved.

Ion exchange is a process already used in the waste water technique.

The applications can be summarized as follows:

(1) As tertiary treatment of waste water to remove phosphates. This means that the process replaces chemical precipitation. The cost is indicated to be 0.02 $ m^{-3} including the cost of operation, amortisation of capital investment over a 10-year period, and the resin over a 3-year period, excluding the cost of disposal of the waste regenerant [28, 30].

(2) As the efficiency is higher than with chemical precipitation and the cost is only slightly greater this process is worthwhile considering as an alternative solution.

TABLE II

A survey of the application for the ion exchange process for treatment of municipal waste water

Type of waste water	Components removed	Ref. no.
Metal Plating and Finishing Industry	Several metal ions incl. chromate	[33] and [37]
Iron Industry and Mining	Iron	[16]
Electrolytical Industry	Mercury	[6]
Manufacturing of Glass- and Stone wool	Phenol	[21]
Fertilizer Industry	Phosphate, Ammonium, Nitrate	[49]
Food Industry	Proteine, Carbohydrates	[22]
Textile Industry	Dyes, Organic chemicals	[23]

(3) As a demineralisation of tertiary sewage effluent to produce water acceptable for industrial use. Organic as well as inorganic content is reduced 3 to a very low level. The cost for a plant treating 4580 m^3 per day is approximately 180000 $ and the running costs are 0.15 $ m^{-3} [7].

(4) Table II gives a survey of the application of the ion exchange process in the treatment of industrial waste water.

As a conclusion of the above-mentioned considerations it is quite obvious that this unit operation has a wide range of applications within the waste water treatment field.

When the process has not found an even wider use it is due to:

(a) the more polluted the water, the more easily the ion exchanger will clog due to fouling. The macroporous ion exchanger has reduced but not completely solved this problem [26],

(b) the efficiency of removal of organic matter is in most cases not high enough when very polluted water is treated by this process,

(c) the operation cost is proportional to the frequency of regeneration. Highly polluted water requires more frequent regeneration and the cost of treating such waste water is therefore quite high.

From these considerations it is easy to understand that the process has found its main application in the treatment of moderately polluted water. The capacity of the ion exchanger is determinative whether the process is able to compete with ot heralternative solutions to the same problem.

3. The Combination of Precipitation and Ion Exchange

The combination of the 2 unit processes mentioned – precipitation and ion exchange – seems to offer many advantages, since the precipitation process is more suitable for

PROCESS SELECTION AND OPTIMISATION FOR WATER POLLUTION CONTROL IN THE PULP INDUSTRY

E. NORMAN WESTERBERG

EKONO, Helsinki, Finland

Abstract. From studies over the interplay between plant and recipient water, characteristics of pulp and paper mills and process costs, a 'System Design and Optimisation Program for the Chemicals Recovery Cycle' (SYDOP-CR) is built up and exemplified.

Traditionally, the planning, construction, and operation of industrial plants has involved selection of proper mill size and processes considering available raw materials and markets, with the usual criteria on a venture's efficiency being its profitability measured as net return on its capital investment. Today, environmental control considerations enter as restraints in this optimisation program. In the battle between ecologists and economists on one hand, and between regulatory agencies and industry on the other, I see the engineer's role as one of supplying quantitative information on the interplay between the plant and the environment. This means interpreting legal paragraphs and replacing the many question marks in Figure 1 with dollar signs as far as possible.

Both as industrial consultants and as regional planners, the author's organisation has always favored a systems engineering approach to problem solving. This paper discusses some aspects of this approach and can be seen as consisting of two parts: (1) a general discussion of the environmental protection problems involved and (2) a description of some of the optimisation programs for process and equipment selection that have been developed, with application examples from the pulp and paper industry.

In most instances there are many industries, cities, and municipalities located along a water course, drawing benefits from the many uses of water and discharging effluents in varying amounts and with varying characteristics. Ideally, one would like to have a mathematical model of the entire system by which one could calculate or predict the effect anywhere in the recipient water of the change in an effluent anywhere else in the system. This may seem as a utopia, but it is the author's contention that the development of such models must be our goal and, that in many instances, it is already today possible to develop and use at least simple models usefully to a much greater extent than is being done. The chemical engineer, working closely together with engineers and scientists of other disciplines, will play an ever increasing role in this development.

1. Interplay Between Plant and Recipient Water

In their simplest form the systems discussed here can be broken in 4 blocks as shown in Figure 2. I have on purpose shown a block 'in-plant recovery or treatment' separate from the production plant block, although in the pulp and paper industry this in-plant

G. Lindner and K. Nyberg (eds.), Environmental Engineering, 251–262. All Rights Reserved

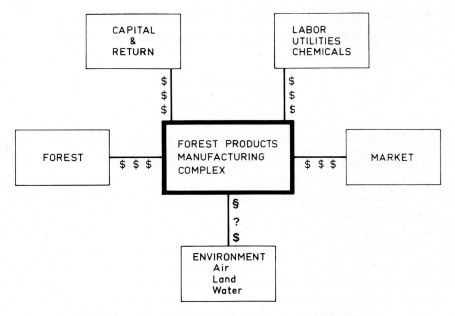

Fig. 1.   Effluent control – a part of overall planning and economics.

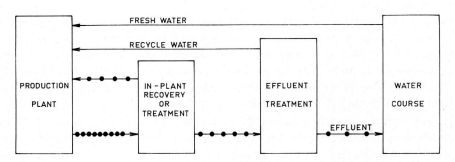

Fig. 2.   Basic process blocks in effluent-water cycle.

treatment already is standard procedure to a very high degree. There is still considerable room for improvement, however, and this in-plant treatment is a prerequisite to
successful so-called outside treatment, shown in the third block.

There seems to be no dispute about the fact that regulatory requirements will be
necessary in order to induce industry to reduce its effluent amounts. On the other hand,
it must be improper, at least in the long run, to express such requirements in terms of
so many kilograms of BOD per ton of product, or so many tons of BOD per day from
a given plant, without due consideration of the local circumstances and the natural
capabilities of the recipient waters for purification. With limited economic resources,
in an optimum system, the money should be spent where it does the most good, and,

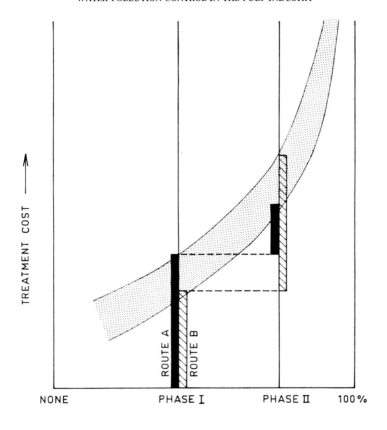

Fig. 3.   Diagram illustrating a problem of route-selection with step-wise
implemention of treatment facilities.

where necessary, a system for sharing the cost must be worked out. It is of course the
non-availability of proper models for predicting the behaviour and response of the
recipient itself, the fourth block in Figure 2, that makes it necessary for regulatory
agencies at the present stage to use such measures.

Another problem that industry is faced with is the so-called moving target, illustrated in Figure 3. The shaded area may represent a hypothetical curve of treatment
costs vs the degree of treatment. Assume that at a certain time the requirements for
treatment correspond to phase I in the diagram. The plant can choose between two
treatment processes of different costs, represented by columns A and B respectively.
Some time later, the treatment requirements will be increased to point II in the diagram, and the cost of additional treatment again represented by the columns shown.
Obviously, if at the first phase a later increased requirement is disregarded, the plant
would choose the treatment method B as being less costly than A. On the other hand,

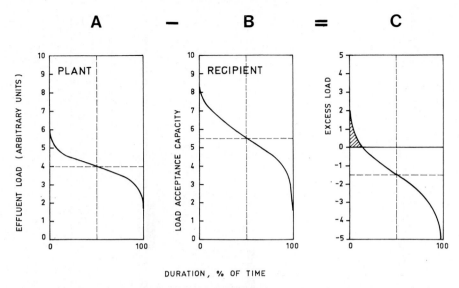

DURATION, % OF TIME

Fig. 4.   Calculation of excess load duration based on variation in load and
recipient's capacity receiving load.

if the increased requirements can be foreseen and are due in a reasonably short time
the plant may select the A route as giving the lowest final treatment cost. In my opinion, in many cases the step-wise regulatory requirements have been rather arbitrary
and the time intervals too short for industry to really study and develop proper treatment methods and long range integrated systems.

Another point of interest is illustrated in Figure 4. Let diagram A represent typical
variations in the amount of some pollutant and let diagram B represent the corresponding variation in the recipient's capability of accepting this pollutant. The variations in diagram B could for example be due to fluctuations in river flow and temperature affecting the recipient's total availability of oxygen for biological degradation.
Diagram C may be said to show the excessive load and the available excess purification
capacity respectively, on the assumption that the variation in Diagram A and B are
completely independent of each other. Now, assuming that the excess load cannot be
tolerated, the plant has at least the following three alternative choices:

(1) To reduce production, and thereby the effluent load, during those days when
an excess occurs.

(2) To build a hold-up basin to store the effluent during days of excess load for
later release.

(3) To increase the efficiency of the treatment.

2. Some Characteristics of Pulp and Paper Mills

Figure 5 shows a simplified block diagram of a pulp and paper mill. Such mills have
the following characteristics:

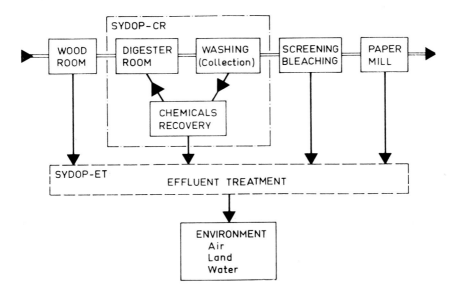

Fig. 5.   Simplified diagram showing boundaries of SYDOP programs.

(1) They have a high water consumption, normally in the range of 200–600 m^3 ton^{-1} of product.

(2) They have an intrinsic high BOD load, due to the fact that in the cooking process about half of the wood substance is dissolved in the chemical cooking solution.

(3) Typical waste loads for today's pulp mills, having a reasonable degree of spent liquor recovery, are:

	Kg/ton pulp
Suspended solids	10 –100
BOD$_5$	30 –100
Nitrogen (N)	0.4– 0.8
Phosphorus (P)	0.6– 1.5

Fortunately, with regard to (2), for most processes suitable methods have been developed for destruction of the organic effluent and recovery and recirculation of cooking chemicals, in some instances dictated by economics alone.

The older process, the sulfite process, was first introduced with calcium as a low cost cooking base. In the early 1950's suitable methods were developed and rapidly introduced in Sweden and Finland for evaporation and burning of the spent cooking liquor, thereby recovering at least the heat value of the spent liquor. At about the same time methods were also being developed for evaporation and burning of spent liquors from magnesium and sodium based sulfite pulping processes, where in addition also the chemicals could be recovered for reuse in the process. Also ammonia as a base chemical has found use, particularly in the U.S.A.

The other, and today much more wide-spread pulping process, the sulfate or kraft process, was primarily developed to find a process to pulp a wider variety of wood species (pine, etc.). Since it is an alkaline process, cheaper construction materials (carbon steel) could be used more widely in the process instead of stainless steel necessary in the sulfite process.

In the more tricky field of sulfite recovery, Finland leads in both relative and absolute terms in the amount of sulfite spent liquor recovered.

Today a pulp mill faced with the problem of spent cooking liquor disposal or recovery has a large variety of alternatives to choose between. A systematic approach to comparison of these alternatives, and to evaluation and storage of operating and cost experience, was therefore logical, and led to the development by the author's organisation of what we call a 'System Design and Optimization Program (SYDOP-CR)' for the chemicals recovery cycle. The process boundaries of this program are shown schematically in Figure 5.

In previous years, at least in Finland, optimum design parameters for recovery installations were selected primarily on chemicals and fuel economy considerations alone. Today, the resulting degree of spent liquor solids collection may not, and in many instances will not, fulfill the requirements posed on the system by the environment. It is recognized, however, that the methods and degree of any further treatment of uncollected spent liquor, as well as of new effluents stemming from the recovery system such as dirty condensates and odorous gases, must be viewed along with the treatment demands and possibilities of effluents from other mill departments (wood room, bleach plant, paper mill, etc.). A separate SYDOP-ET for the effluent treatment parts is, therefore, being developed, based on exactly the same principles and programming techniques as that for the chemicals recovery cycle.

3. Objectives and Description of SYDOP-CR

The main objective of the chemicals cycle program is to generate investment and operating cost data for any feasible combination of process units, equipment and operating parameters. In the course of this, the program also develops materials and heat balance information, including amounts and characteristics of solid, liquid, and gaseous effluents.

A second objective is to free experienced engineers from manual routine calculations, allowing them to use their ingenuity for proposing technically feasible process combinations and for evaluating the comparative economies of alternative solutions.

A third objective is to provide a consistent frame-work for evaluation and storage of operating and cost experience, both existing and yet to be gained. Most of the process units and alternatives listed in Table I have been programmed and tested in many applications.

TABLE I

Typical process blocks and their application, SYDOP-CR

Process block	Typical alternatives or examples
Cooking	Continuous, Batch
Blow	Aux. block serving as intermediate between cooking and collection
Collection (washing)	Blow pits, Rotary, Diffusion, Hi-heat, Presses
Liquor treatment	By-products (alcohol, yeast, soap). Oxidation (kraft), Steam stripping, Reversed osmosis, Ion exchange, Electro-dialysis:, strong or weak liquor
Evaporation (indirect)	M.E., Recompression, Flash evaporation
Evaporation (gas contact)	Strong or weak liquor
Liquor burning	Reductive, Oxidative, Fluidized bed, Pyrolysis
Auxiliary furnaces	SCA-Billerud
Steam generation	Different pressure and temp. levels
Power generation	
Ash separation	Electr. precipitator, Multiclones, Wet scrubber
Ash (or smelt) handling	Leaching (SCA-Billerud), Smelt oxidation or carbonation, Green liquor preparation
Flue gas cooling	⎫
SO₂ absorption (flue gas)	⎬ Venturi, Spray towers, Packed towers, TCA
Fortification	⎭
Chemicals conversion	Institute, Sivola, Stora, Tampella, etc.
Causticizing	
Clarification	Acid polish., Clarification and mud wash.
Lime mud reburning	Rotary kiln, Fluidized calciner
Base preparation	e.g. Soda ash dissolving
SO₂ preparation	Burning of sulfur and/or sulfurous gases
Acid system	
Condensate treatment	Stripping, Reversed osmosis
Odorous gas treatment	Combustion, Absorption

Each major process unit can be schematically represented by one block, which may be connected to other blocks using at the most ten flows.

Each flow can be characterized by ten flow parameters, such as total flow, amounts of chemicals, temperatures, etc. The example in Table II is typical for all spent sulfite liquor flows.

In addition, for each block a maximum of ten block parameters may be defined, e.g. liquor displacement performance characteristics in the collection block.

A sub-routine, describing a process unit consists of equations for heat and material balances for the unit i.e. it tells how the output flows are determined from input flows and block parameters. The connection of process units to each other is determined by the configuration matrix, and one row of this matrix must be determined for each block.

The printed flow matrix consists of 50 rows and 10 columns. Each row number refers to a flow and each column refers to a specific characteristic of the flow for the corresponding row.

In addition to the flow matrix, which is printed out for individual runs only when requested, we normally use the following two print-out formats:

TABLE II

An example of a flow, typical for all spent sulfite liquor flows

Parameter No.	Characteristic	Unit
(1)	Total flow	lb tp^{-1}
(2)	Temperature	deg. F
(3)	Base amount (e.g. MgO)	lb tp^{-1}
(4)	Bound sulphur (as SO_2)	lb tp^{-1}
(5)	Non-volatile wood substance (WS)	lb tp^{-1}
(6)	Sulfate (as SO_3)	lb tp^{-1}
(7)	Volatile organics (VO)	lb tp^{-1}
(8)	Free or loosely bound SO_2	lb tp^{-1}
(9)	Water (H_2O)	lb tp^{-1}
(10)	Enthalpy of liquor	B.t.u. lb^{-1}

TABLE III

Illustration of individual run data print-out

SYDOP for MGO process Faasic project, investment Oct. 1969 million Can. dollars, production 10.0 tp hr^{-1}, run No. 1

	Investment	Consumption		Generation
		Steam MMBTU TP^{-1}	Power kWh TP^{-1}	Hot water MMBTU TP^{-1}
Digesters	0.00	3.2	36.0	0.0
Collection	0.73	0.0	34.0	0.5
Evaporation	1.75	4.5	42.0	4.5
Combustion	1.05	1.3	29.0	0.0
Steam generation	0.82	0.3	13.0	0.0
Ash separation	0.18	0.0	7.0	0.0
SO_2 absorption	0.87	0.0	58.0	4.6
Base preparation	0.33	0.2	12.0	0.0
S-burning	0.00	0.0	5.0	0.2
Fortification	0.07	0.0	5.0	0.0
Acid system	0.00	0.0	19.0	0.2
Power generation	0.00	0.0	0.0	0.0
Other	0.20	− 4.0	− 60.0	0.0
Total	6.00	5.5	200.0	10.0

Make-up:
  – Base   20.0 lb TP^{-1}
  – S   60.0 lb TP^{-1}
Aux. fuel used   0.0 MMBTU TP^{-1}
Steam generation   14.5 MMBTU TP^{-1}
Power generation   0.0 kWh TP^{-1}

(1) A tabulation for each run or case showing investment costs, steam and power consumption, and hot water consumption and generation, for each process block as well as the total system, plus data on demand for make-up chemicals, steam and power generation, etc. An illustration of such a tabulation, somewhat abbreviated and with rounded-off numbers, is given in Table III.

NEUTRALIZATION OF INDUSTRIAL WASTE WATERS BY LIME

RUDOLF RYCHLÝ

Research Institute of Inorganic Chemistry, Ústi nad Labem, Czechoslovaki

Abstract. Lime is by far the most widely used neutralizing agent for industrial acidic waste waters. It is important, however, that the by-product, i.e. gypsum, can be used too. On laboratory experiment results with crystallization of gypsum, the chemical and physical parameters for an ideal by-product formation is discussed.

The ever increasing requirements of the Czechoslovak water management and the responsible authorities represent a very complex task and challenge for chemists and chemical engineers, especially in the field of purification of acid industrial waste waters. There are, for instance, several reports (Bahenský, 1969; Malyk, 1966; Petlička and Bástečký, 1964), dealing with the neutralization of spent steel pickling liquors by lime (which has been the most widely used neutralizing agent so far), and with the consequent problems of reducing the volumes of neutralizing sludges and further processing and utilization of gypsum.

1. By-Product Problems

A Czech plant producing titanium dioxide may be cited as an extreme source of liquid wastes from chemical industries. The composition of these waste waters fluctuates substantially (SO_4^{2-} : 2–50 g l^{-1}; Fe: 0.6–14 g l^{-1}; soluble TiO_2: 0.06 to 3 g l^{-1}) and this results in a very heterogeneous neutralization sludge, which is the reason why thus by-produced gypsum cannot be further utilized. The disposal of large volumes of the neutralization sludge obtained by the treatment of some 200 m³ h^{-1} of acid waters presents serious troubles. Another serious difficulty is the formation of scale on the whole piping system.

Those processors who might be interested in the recovery of the waste gypsum from lime-milk neutralization processes, e.g. by its conversion to a building industry material have set rather stringent standards concerning the quality of gypsum crystals and the content of impurities, for instance, when production of plaster is to be carried out, table-shaped gypsum crystals are preferred. This and also the questions of sludge filterability (separation) as well as the scale formation make it necessary to study profoundly the complex problems of gypsum crystallization during the neutralization of waste waters by lime.

The methods of experimental work on Czechoslovak through-flow waste treatment units had been developed and described in detail earlier (Dvořák, 1960, 1965; Tesař, 1960). The problems of scale-formation (Dvořák *et al.*, 1969, 1970) and gypsum crystals size and shape (Tesař, 1971; Dvořák and Rychlý, 1971, 1972) have been given due attention only recently.

2. Dissolution and Crystallisation of Gypsum

The dissolution of gypsum between 10°C and 30°C follows (Konak, 1971; Sung-Tsuen and Nancollas, 1971) an equation first order in relative subsaturation and is sensitive to changes in the rate of stirring indicating film diffusion control. The rate of dissolution under similar stirring conditions were some fifty times faster (Smith and Sweett, 1971) than the crystallisation rates, showing the kinetic importance of surface processes in the crystallisation. The crystallisation on the addition of seed gypsum crystals to stable supersaturated solutions (prepared, e.g. by mixing calcium chloride and sodium sulphate solutions) has been studied chiefly (Konak, 1971; Sung-Tsuen and Nancollas, 1970). The crystal growth follows (Smith and Sweet, 1971; Sung-Tsuen and Nancollas, 1970) an equation second order in relative supersaturation and the rate constant is independent of surface area changes. The value of the rate constant indicated (Smith and Sweet, 1971) that the rate controlling step was dehydration of calcium ions.

Results of unseeded experiments (a dilatometric study at 30°C) suggested (Smith and Sweet, 1971) that nucleation was heterogeneous. The effect of agitation was to increase the nucleation rate, thus suggesting (Konak, 1971) a partially mass-transfer controlled process for nucleation.

3. Laboratory Experiments

Laboratory experiments (discontinuous and continuous) were done and the difference between recrystallisation of gypsum in HNO_3, H_3PO_4 and sulphuric acid medium was

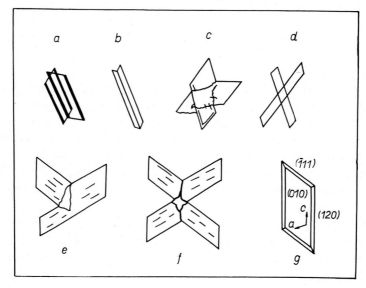

Fig. 1.   Gypsum morphology according to Simon (1968).

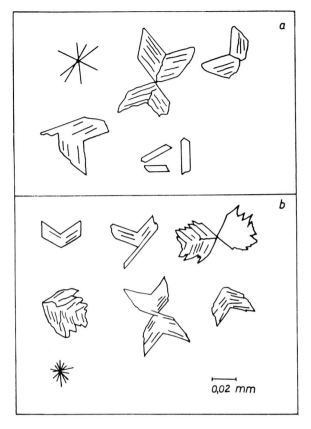

$$CaSO_4 \cdot 2H_2O$$

Fig. 2.   Gypsum crystals: (a) precipitation from a dilute nitric acid, (b) precipitation by
aqueous suspension of $Ca(OH)_2$.

pointed out, and formation of needle crystals, radially arranged clusters of the needle-
crystals, as well as the V- and X-shaped twins ($\bar{1}01$) and overgrowing of some aggregates to the twins ($\bar{1}01$) was observed. It was found that the effectiveness of the crystallization decreases with an increase in the pH of the sludge. The degree of stirring
has almost no effect on crystallization. At higher stirring intensities the growth rate
of gypsum crystals reaches a certain constant value. A difference was found between
discontinuous and continuous experiments on lime-stone neutralization. (Dvořák and
Tesař, 1971). In the medium of pH 1.5 there were more needle-crystals than twins at
pH 2 to 3.

The morphology of gypsum in aqueous solutions according to Simon (1968) is
shown in Figure 1. With regard to the separation of gypsum and its subsequent treatment the most advantageous morphology is that of single crystal (Figure 1g).

The shape and size of gypsum crystals obtained either from solutions of dilute sulphuric acid ($10-70$ g l^{-1} H_2SO_4) or from model samples of waste water (37 g l^{-1} H_2SO_4, 3 g l^{-1} Fe, 0.4 g l^{-1} TiO_2) by neutralizing the respective solutions by aqueous suspensions of calcium hydroxide ($3-50$ wt.%) is demonstrated in Figure 2, the volume of reaction mixtures always being 300 ml (Figure 2b). For comparison gypsum was also precipitated by mixing the solution of calcium nitrate (7.5 g $Ca/NO_3/_2 \cdot 4$ H_2O per 100 ml H_2O) with sulphuric acid, 34 g l^{-1} H_2SO_4, in a volume of 200 ml (Figure 2a). In both cases the initial stage of precipitation was characterized by the formation of needle crystals unsuitable for processing to plaster, radially arranged clusters of needle crystals, as well as V- and X-shaped twin crystals ($\bar{1}01$), so called 'swallows'. The overgrowing of some radiate aggregates to the twins ($\bar{1}01$) was also observed.

The experimental conditions were as follows: the reaction mixtures were stirred at $300-500$ rpm for 120 min, then their agitating was discontinued and the samples were kept undisturbed at ambient temperature. The occurrence rate of the individual crystal forms was evaluated after 3 h. The sample obtained by neutralizing the model waste water by a 10% suspension of $Ca(OH)_2$ to pH $= 3.5$ contained about 50% of V- and X-shaped crystals, 28% needle crystals, 7% table-like crystals with single crystal morphology, the remainder being radiate clusters of needle crystals overgrowing to ($\bar{1}01$) and ($\bar{2}01$) twins. One of the samples was kept undisturbed for a period of two months at room temperature, pH 2.6. The occurrence of V- and X-shaped twin crystals ($\bar{1}01$) was then prevailing, though a more detailed microscopic examination of a greater number of crystals revealed the presence of a substantial portion of needle-like crystals that were fairly well developed with single crystal morphology. The number of needle crystals was unambiguously prevailing in a neutralization sludge sample kept at pH $= 8.6$ for two months.

The situation, however, was somewhat different with the recrystallisation of gypsum precipitate in dilute nitric acid obtained by mixing solutions of sulphuric acid and calcium nitrate, pH $= 1$. The gypsum crystals shape was tabular, and exhibited transition to single crystal morphology. In addition, V-shaped twins ($\bar{1}01$) were present, while fine needle-like crystals were virtually absent.

The rate of needle crystals occurrence is increasing with the value of pH. It is thus possible to draw a general conclusion that bigger gypsum crystals, and consequently less cohesive scale, are formed during the crystallization and even more during recrystallization stage in the acidic region of pH. The growth rate was found to decrease slightly with increasing pH. No significant influence of the method and rate of both combining the reactants and subsequent agitation of the resulting slurry on the shape and size of gypsum crystals could be detected.

4. Large-Scale Experiments

The above results compare well with the results of experiments conducted on a bench-scale through-flow model of 2-stage neutralization, first stage neutralization to pH $=$ $= 3.5$ in a 1.5 l vessel, second stage neutralization to pH $= 10$ in a 15 l vessel equipped with an agitator (Dvořák *et al.*, 1969, 1970; Dvořák and Tesař, 1971). The treated

TABLE II

The waste water running-in expressed
through the mean value and
the standard deviation

pH	7.8 ± 0.2
alkalinity	7.60 ± 0.32
turbidity	57 ± 8.1
COD	304 ± 55
o-phosphate	9.00 ± 2.00
t-phosphate	12.50 ± 2.30

TABLE III

Inflow					Outflow			
$o - P$	$t - P$	COD	Turbidity	pH	$o - P$	$t - P$	COD	Turbidity
8.80	12.30	332	69	8.9	2.76	4.14	189	38
7.90	11.80	304	58	9.2	2.56	3.96	185	36
7.35	9.50	237	49	9.3	2.98	4.10	148	31
10.25	12.40	307	67	9.4	2.88	3.50	132	27
9.75	13.20	282	62	9.4	2.98	3.95	155	32
8.53	12.50	291	63	9.4	2.24	3.95	168	28
9.80	16.60	400	82	9.4	1.44	3.75	164	27
10.20	16.90	375	69	9.4	2.03	3.75	195	40
10.05	16.50	380	72	9.5	2.14	4.75	212	47
10.25	16.00	363	69	9.5	1.66	3.65	186	36
8.13	11.00	417	64	9.8	1.05	1.35	146	18
8.95	10.80	337	60	10.1	0.98	1.35	148	20
9.70	12.20	375	59	10.1	1.02	1.48	145	20
10.25	11.70	265	48	9.5	1.70	2.59	129	26
9.56	13.80	302	53	9.1	1.44	2.47	138	21
9.90	12.00	303	59	9.7	2.25	3.50	142	26
12.80	16.80	312	56	9.7	2.50	2.70	171	27
3.90	10.80	254	46	9.4	0.78	2.63	151	37
4.00	12.00	297	51	10.1	0.53	1.63	141	22
4.40	13.10	297	52	10.2	0.48	1.40	148	18
3.20	12.80	294	52	10.5	0.37	0.75	116	14
10.70	12.40	241	46	11.0	0.17	0.34	121	4
10.70	13.80	278	56	11.1	0.17	0.35	112	5
9.90	13.20	227	48	10.8	0.11	0.33	96	7
9.60	13.80	254	51	10.4	0.44	0.99	119	13
7.90	10.00	216	47	11.3	0.14	0.51	72	5
9.00	11.40	243	54	11.4	0.20	0.35	124	3
10.00	14.00	296	61	11.4	0.09	0.30	90	6
9.00	16.00	277	60	11.4	0.27	0.64	107	15
9.60	9.60	253	51	9.7	1.44	1.79	109	18
8.80	9.00	223	52	9.7	1.54	2.00	146	22
9.10	9.50	303	61	9.7	1.68	2.33	149	24
9.90	11.00	251	44	9.8	0.75	1.33	93	16
11.80	14.30	389	63	10.1	0.52	1.03	81	13
11.00	16.00	219	45	10.2	0.60	1.19	89	17

METHODS AND PROCESSES FOR
WASTE WATER PURIFICATION IN THE FOOD INDUSTRY

BØRGE F. MORTENSEN

F L Smidth & Co A/S, Environmental Technology Division, Copenhagen, Denmark

Abstract. In Denmark the food industry is a comparatively big polluter This is, of course, due to the production profile of the country. Special attention must also be paid to those specific methods and often advanced techniques that is the consequence of such an industrial activity. Methods and purification models are accounted for.

1. Food Industry as a Source of Pollution

From the point of view of industrial effluent problems, the food processing industry counts heavily in Denmark. This is in contrast to a number of other European countries where other industries contribute a major part of the industrial water pollution load, e.g. pulp and paper, metallurgy, etc.

A recent survey by the Danish Pollution Control Council uncovered the contribution shown in Table II. The values may be summarized as shown in Table I.

It appears that while the water consumption is about one fourth of the consumption by the households, the pollution load nearly equals that of the households.

The treatment costs will be high even when only traditional mechanical and biological methods are used and will rise even higher if and when phosphate or nitrogen removal is required. These facts should invite for new treatment methods based on our specific knowledge about the concentration and chemical composition of the food industry effluents.

Is it at all possible to generalize when we know that a great variation exists, not only between the various branches of the food industry, but also between industries of the same type? Even for the same industry there are great variations from year to year, from week to week, from day to day, and even from hour to hour. Seasonal variations are for instance important for the beet sugar and potato processing plants and for the fruit and vegetable industry. On the other hand, slaughtering operations now show a much more uniform day to day picture due to an industry-wide modernisation and mechanisation. Still, it is believed that a generalised picture of the available treatment methods and their efficiency may be useful.

2. Classification of Pollutants

Common to all of the food industry are effluents containing biodegradable matter as the main pollutant load. This makes traditional biological treatment, for instance in a municipal plant, easy to apply in theory. In practice, difficulties arise due to large variations in the load of such a plant. In fact one must either install a large capacity

TABLE I

	Water consumption million meter³ day⁻¹	Pollution load in million PE[a]
Meat Industry, total	21.3	0.88
Milk Industry, total	27.1	0.35
Vegetable Industry, total	14.5	1.59
Fermentation, total	8.0	0.50
Miscellaneous, total	1.2	0.12
Food Industry, total	72.1	3.44
Households, total	340	4.12

[a] PE = person equivalents – 60 g BOD_5 day⁻¹.

TABLE II

Water consumption and pollution load by food industry in Denmark

Industry	Water consumption unit: million meter³ day⁻¹	Pollution load in million PE	Number of working days per year
Abattoirs	13.7	0.60	250
Meat-Canning Industries	4.4	0.24	250
Renderers	3.2	0.04	250
Meat Industries total	21.3	0.88	–
Dairies	18.4	0.27	360
Milk Canning Industries	8.7	0.08	250
Milk Industry total	27.1	0.35	–
Fruit and Vegetable Industries	1.5	0.14	250
Sugar Refineries	10.0	1.0	60–80
Potato Starch Industries	3.0	0.45	120
Vegetable Industries total	14.5	1.59	–
Distilleries	0.2	0.10	250
Yeast Producers	0.8	0.18	250
Breweries	5.4	0.18	250
Malt Houses	1.6	0.04	250
Fermentation Industry total	8.0	0.50	–
Soft Drink Industry	1.2	0.11	250
Margarine Industry		0.01	250
Miscellaneous total	> 1.2	0.12	–
Food Industry total	> 72.1	> 3.44	–
Households total	340	4.12	365

which cannot be fully utilised at all times, or one must accept overloadings from time to time.

This unfortunate situation may induce industry to install pre-treatment systems or to build its own plants treating to required standards, but using new and/or special methods. In order to facilitate the discussion of the various possibilities, an attempt has been made to classify the pollutants according to their chemical composition and physical form. This is illustrated in Table III.

TABLE III

Pollutants from the food processing

Composition	Physical form	
	In solution	Suspended or emulsified
Major constituents		
Microbial cells and debris		X
Proteins	X	X
Fats and oils		X
Carbohydrates	X	X
(Nutrients)	(X)	
Minor constituents		
Acids/bases	X	
Salts	X	
Detergents	X	
Desinfecting agents	X	
Food additives	X	
Grit		X

3. Physical Treatment

The non-dissolved pollutants can often be removed by physical means. Solids may be screened off by use of self-cleaning sieves of various constructions (rotating or vibrating types are common). Removal of fine solids requires a retention time of 1–2 hr in a sedimentation unit or about 15 min in a dissolved air flotation unit. For some industries grit removal is important to prevent excessive wear of machinery. Grit is removed in grit traps similar to those found in municipal plants.

Fats and oils have a specific gravity less than water and rise to the surface in a short time except when very finely dispersed. For heavy loadings, a grease trap should be cleaned often, and automatically if possible. Removal of fat can also take place in centrifugal separators or by air flotation.

4. Biological Treatment

The dissolved pollutants consist mainly of proteins and their products of decomposition, i.e. peptones, polypeptides, amino acids and carbohydrates. They are readily bio-oxidised by bacteria in a trickling filter or an activated sludge plant, provided the necessary nutrients are available. If not, they should be added. Traditional biological treatment requires a large area and volume. Today, the plastic trickling filter media make possible the use of tall and light constructions with a much higher loading than conventional and with a consequent decrease in area requirements.

Organic wastes with BOD_3 over 2000–3000 ppm may be decomposed quickly by anaerobic digestion, a method which is used for the treatment of yeast and pectin effluents in Denmark, and also for slaughterhouse, mezt-packing and brewery waste-water in the U.K., U.S.A., and New Zealand.

Inexpensive and often very efficient is lagooning, where both aerobic and anaerobic processes oxidise and decompose the organic matter during 20–30 days of retention time in a pond system. Drawbacks are land requirements and the low conversion rates during cold seasons. For these same reasons land treatment by irrigation and soil oxidation is only used by isolated plants in less populous areas.

5. Chemical Treatment

As the proteins are less soluble at their isoelectric point, it is possible to remove part of the dissolved proteins as a sludge by addition of sulphuric acid or acid salts. By further addition of chemicals, such as sulphite lye or pure lignosulphonic acid, non-soluble chemical compounds with protein are formed, and a substantial part of the BOD_5 may be precipitated in this manner. In combination with dissolved air flotation, this process requires very little space.

By addition of aluminium sulphate or lime, flocs are formed, and the phosphates and a majority of the finely dispersed organic particles are precipitated (up to 60% of the BOD_5).

6. Chemicals in the Waste Water

Food industry effluents also contain minor amounts of detergents, desinfecting agents, dyes, salts, and remnants of the processing chemicals used to make the food more attractive, non-perishable or to make processing cheaper. Many of these compounds pass straight through the effluent treatment system because they are only slowly oxi-dised, or not oxidised at all. Consequently, if these pollutants must be removed, it may be necessary to employ some of the more advanced treatment methods.

7. Advanced Treatment

Organic chemicals may be adsorbed on activated carbon or removed by ion exchange or reverse osmosis. One of the most promising of the advanced treatment methods is the use of activated carbon for removal by adsorption of soluble organic pollutants. Other methods such as ion exchange and reverse osmosis, may be used for concentrating the soluble pollutants and thus removing them from the major part of the waste water. These methods are quite expensive, but their economy may be improved if the concentrated pollutants or the treated waste water can be reused.

8. Desinfection Methods

Only a few food industries produce effluents containing pathogenic bacteria. Rendering plant effluents must be pressure cooked for one hour at 115°C according to Danish regulations. Often, chlorination is used for desinfection after biological treatment. To

THE PROS AND CONS OF UTILIZATION OF COMPUTER
CONTROL WITHIN THE POLLUTION CONTROL FIELD

PHILIP MADSEN

The Madsen Co., Los Angeles, Calif., U.S.A.

Abstract. Two types of water handling control systems are exemplified and discussed The central computer with a supervisory scanner is an effective system regarding the simplicity in handling, cost efficiency, adaptability to manual control but will not operate in a stand alone configuration and is not as fast as the computer-to-computer system, in which the advantages and drawbacks are almost the reverse to the other system. Therefore, both systems are in operation and chosen with regard to these factors mentioned

During this presentation I will discuss two basic concepts in water handling control. The first is a control system in which the control center communicates with, logs data from, and controls the operation of, remote stations – each employing a supervisory control scanner. In the second approach, all the aforementioned functions are performed; however, a computer is used at each remote site, as opposed to a supervisory controller or scanner.

1. Computer-Scanner System

The computer-to-scanner concept is shown in Figure 1, which is a simplified block diagram of a supervisory control project being developed for the City of Riverside. In this system, a computerized control center transmits and receives data via nine separate telephone circuits. Each phone circuit consists of a transmit and receive line with a number of station drops. This system has a total of 20 remote stations currently under development and will be expanded at a later time to encompass 50 remote stations. The supervisory controller to be installed on the Riverside project uses large scale integrated circuits for its transmitter and receiver sections. The transmitter has an LSI circuit module with an equivalency of 1500 discreet transistors. The receiver has an equivalency of 2500 transistors. The supervisory controller using the LSI circuits approach has such a high reliability that a 5-year warranty is provided with each scanner.

As a point of interest, the LSI chip is less than 1/16th of a square inch. All inputs to the supervisory controller from the equipment being monitored or controlled are provided with transient protection of up to 1500 volts.

Turning now to the control center (Figure 2) I will briefly discuss the equipment used in this system, since it is fairly representative of numerous other control systems. On the top left of the block diagram (Figure 2) a typical remote station is illustrated. This typical station is shown as receiving a signal from the upper line coupler and feeding to the lower line coupler, both of which communicate serially with the supervisory receiver and transmitter, which in turn interfaces to the central processing unit. The supervisory receiver and transmitter are both basically parallel to serial converters

G. Lindner and K. Nyberg (eds.), Environmental Engineering, 295–300. All Rights Reserved

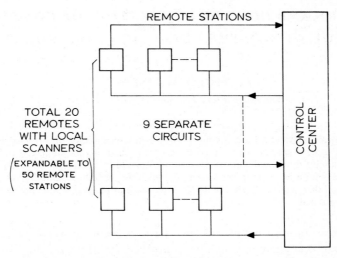

Fig. 1.   Riverside project uses central to control remote scanners.

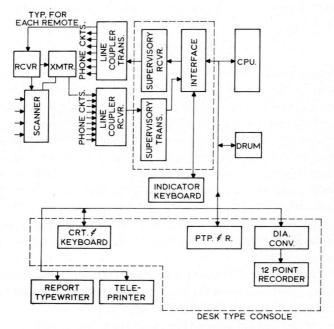

Fig. 2.   City of Riverside control systems.

and security code generators or checkers, secure from the standpoint of low error rate. An indicator keyboard is provided at the center for manual monitoring and control of the system during any down time of the CPU. Standard peripherals are used with this system. The CRT keyboard is the main control and monitoring point. A teleprinter is used for alarm and data logging with a Selectric typewriter used as a report type to generate reports. The 12-point recorder is used for trend recording.

The advantages and disadvantages of the computer scanner approach is highlighted in Figure 3. Obviously, a scanner is much simpler and easier to service and maintain

ADVANTAGES	DISADVANTAGES
(1) SIMPLICITY	(1) WILL NOT OPERATE IN A STAND ALONE CONFIGURATION
(2) COST EFFECTIVE	(2) SLOWER SPEED
(3) READILY ADAPTABLE TO MANUAL CONTROL	
(4) HIGHLY MODULAR	

Fig. 3.   Computer to scanner.

than a computer. It is more cost effective than using a computer at each remote and is very readily adaptable to manual control. It has the disadvantage, however, that it will not operate in a stand-alone configuration, should the communications link be inoperative. The computer-to-scanner approach is limited in speed, however, in comparison to using a computer at the remote site, since the computer may make many scans, store the data, process the data and transmit only the results to the central. The computer, of course, may take direct action at a remote without waiting for control from the central computer. Using the scanner approach, the speed of operation is basically limited by the band width of the transmission medium. In most cases, this is a voice grade telephone line.

2. Computer-Computer System

The MWD project illustrated by Figure 4 is typical of a computer control system. This system has a control center consisting of CPU associated peripherals and modems for communicating with the remote sites. The top large block shows the Joseph Jensen Control Center with its primary central control computer system, shown as System 1, and a backup central control system, shown as System 2. Both systems are identical, with the backup system in a standby configuration, receiving all input data, but performing no data logging or control functions. The Joseph Jensen Control Center communicates with computers at four remote stations. Shown below, the Joseph Jensen system is an Eagle Rock Control System, which is essentially identical to Joseph Jensen, except there are eight remote stations.

A typical MWD remote station is shown in Figure 5. A wired logic unit, which is essentially a scanner, operates when the CPU is down, to report the status of equipment normally monitored and controlled by the CPU. This system shows analog and

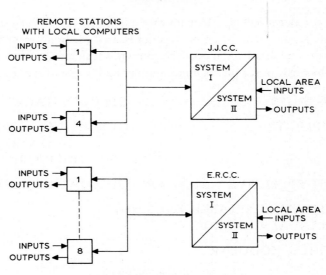

Fig. 4.   MWD project uses centrals to control remote computers.

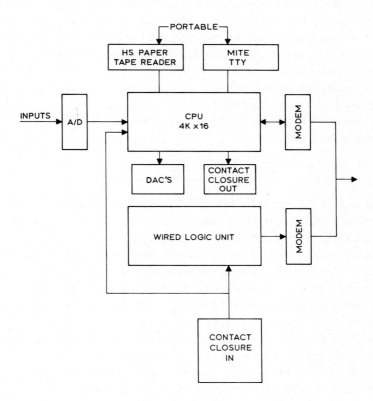

Fig. 5.   MWD remote stations.

3.2.2. *Seres DCO Mètre*

The instrument is housed in a metal cabinet with front and rear doors. Electronic circuits for amplification, temperature regulation and alarm functions are together with electro-mechanical timers mounted on the inside of the front door. Switches and control knobs are only accessable from the inside of the door but results can be read from the outside too, on a door-mounted recorder.

Six single-channel peristaltic pumps are mounted on a rear panel and on a shelf is situated a thermostated block containing six reaction vessels with water cooled reflux condensers. Underneath these is a single beam photometer with flow-cell.

The floor of the cabinet is used for reagent storage and the space available is sufficient for three weeks supply. The instrument is designed for on-stream operation but sample has to be screened through a 50 mesh screen to remove large particles.

Principle of operation: Specific volumes of sample, sulphuric acid, dichromate-solution, mercuric sulphate solution and silver sulphate solution (as a catalyst) are pumped into a mixing chamber by activating the peristaltic pumps for a certain number of revolutions each. The mixture is then fed into one of the six reaction vessels through a distributing valve. It is then heated for 2 hr and finally fed into the flow-cell of the photometer via a second distributing valve, and the concentration of trivalent chromium ions is measured. Thereafter the mixture is drained to waste.

This procedure is repeated every 20 min by using another reaction vessel. In this way one determination is made every 20 min. One of the reaction vessels is, however, normally used for automatic zero adjustments (For data see Table IV.)

TABLE IV

Data of the Seres DCO Mètre

Data:	
Measuring Ranges	0–250 mg O_2 l^{-1}
	0–500 mg O_2 l^{-1}
	0–1.000 mg O_2 l^{-1}
	0–1.500 mg O_2 l^{-1}
	0–2.000 mg O_2 l^{-1}
Reagent Consumption	H_2SO_4: 10 l week^{-1}
	$H_2SO_4/NgSO_4$: 1.5 l week^{-1}
	H_2SO_4-solution: 0.7 l week^{-1}
	$K_2C_2O_7$-solution: 0.7 l week^{-1}
	Distilled water: 1.5 l week^{-1}
Output Signal	0–20 mA or 4–20 mA or 0–100 mV
Dimensions	1.615 mm high × 810 mm wide × 500 mm deep
Weight	Approx. 60 kg
Manufacturer	Seres, BP 297, 13 Aix-en-Provence, France

PART VIII

WATER POLLUTION

Reverse Osmosis

REVERSE OSMOSIS IN PROCESS ENGINEERING

GÖSTA LINDNER

Research Institute of Swedish National Defence, Sundbyberg, Sweden

Abstract. The unit operation principle is handled and reverse osmosis R.O. (hyperfiltration) is taken as an example. Process engineering aspects on the economy of R.O. are discussed and some examples are given especially from waste water purification.

The well-known principle of reverse osmosis has been developed over a decade and many articles have been written about it. Why then write about its general principles? The reason is that the general principles of reverse osmosis as a unit operation and its role in process engineering have not often been discussed.

Reverse osmosis began, as we all know, as a sea-water desalination process. Further studies of the semipermeable membranes immediately showed that high concentrations of salt give high osmotic pressures and therefore the need of high pressure to get water through the membranes. This means that it is easy to damage the membranes, and it is said that a high degree of purification is not possible combined with the need for a high capacity. It is naturally easier with lower salt concentrations and the principle has been applied for brackish water demineralisations, waste treatment systems for municipal wastes and useful processes for waste disposal, and recovery of re-useable or valuable materials in industry. These problems have been treated in several articles and I need only mention the monograph by Sourirajan (1969) in this respect. At our Institute they have been dealt with by the author (Lindner, 1971) and Mårtensson (1971a, b) with the emphasis, however, on problems with suspended solids, acids and other chemicals destroying membranes.

1. Waste Water Purification

Waste treatment processes are designed to remedy public health problems and try to restore the ecological equilibria between man, society, industry and nature by reducing bacterial contamination in receiving waters, to eliminate nuisances by removing odors and unsightly solids, and to protect fish, wildlife and recreational facilities by all substances which in some way act as a poison. The limits of effective treatment by removing scum- and sludge-forming and oxygen-demanding substances, not to forget conventional processes in modern sewage plants have nearly been reached and must be complemented by process engineering with regard to transfer of momentum, heat and mass and also with the principles of modern chemical and biological reaction engineering.

The problem of this limitation is increased because all nations have been tapping their natural supplies of fresh water at a rapidly expanding rate. These problems have also arisen in Scandinavia and the fact that almost everyone is accustomed to be able to swim almost anywhere does not make the situation easier. Over the years there has

G. Lindner and K. Nyberg (eds.), Environmental Engineering, 311–318. All Rights Reserved

been a limited and gradually increasing industrial, agricultural and recreational re-use of treated waste waters to supplement conventional supplies. It has now become clear that waste waters must more and more be considered as a water resource of increasing value. Rapidly increasing urban and industrial growth has focused the attention of the public on the need for more effective methods of treatment of all sorts of wastes. Advanced treatment of our waste waters is thus needed not only to protect the resources we now have and to maintain the high quality of the natural environment, but also to enlarge our supply of clean water for all of our uses. Reverse osmosis appears to be one of the most promising of the general techniques of water purification now under study for application to waste water problems. It offers the possibility of removing organic and inorganic contaminants by one process in a single step. A total or marked reduction in biological contamination can be expected, and the process of subsequent disinfection for a safe water supply can be accomplished with little difficulty. In most cases, a quality of water can probably be produced which will satisfy health and aesthetic requirements and be suitable for all purposes for which a natural surface water supply can be used.

Having now summed up what has been written in various publications about waste water, it is easy for a chemical engineer to see that little of this has a background either in process engineering, or in chemistry. The only points having to do with process engineering are preliminary ideas on recirculation.

Some years ago at our Institute we had the opportunity to search through the literature on waste water and waste water purification, and we found very little in chemical analysis having to do with surface chemistry parameters. Because of purification possibilities with reverse osmosis, the lifetime of the membranes, etc., it is necessary to know more about the properties of the solutions, emulsions or dispersions fed to the unit and there divided into pure water and a concentrate.

We have never seen a discussion of how much may be saved in rationalization by reverse osmosis – fewer steps, larger throughput and so on. The only thing discussed is membrane and energy cost and also the cost of apparatus.

2. Industrial Problems

In industrial applications the tendency is exactly the same. A lot of examples are given, as

(1) concentration of sugar solutions, such as cane sugar, beet juice, maple syrup and corn syrup,

(2) concentration of fruit juices,

(3) concentration of coffee and tea prior to freeze drying,

(4) treatment of acid mineral waters, which contain dilute H_2SO_4 and valuable minerals,

(5) concentration and fractionation of cheese whey,

(6) concentration of pulpmill effluents for incineration,

(7) concentration of photographic developers,

(8) concentration of cultivation solutions for yeast mould a.o.,

(9) recovery of protein and carbon hydrates from starch wash waters,

(10) treatment of electroplating waste waters for recovery and re-use of chemicals and water,

(11) concentration of pharmaceuticals in aqueous media,

(12) petrochemical and petroleum separation processes. No satisfactory method is, however, available for accurately predicting the performance characteristics of a particular solution/membrane system. There is a quantity of data from laboratory-scale units published but not of the kind necessary for process engineering discussions.

3. Reverse Osmosis Apparatus

The feed must be defined so that the temperature, pH, osmotic pressure and viscosity are known, and also the concentration of solutes, the presence of microorganisms (the degree of purification necessary and whether there are micro-organisms destroying the membrane), the presence of suspended solids and their size, form and consistency and at last the chemical reactivity of the feed against the membrane material. The information about the feed is at first used to decide about pretreatment, changing of pH, filtering, etc.

The feed will pass through the unit and for this we have to decide about some chemical engineering parameters as production capacity, flux rate for solvent, degree of separation wanted, the necessary pressure and the flow velocity to minimize concentration polarization. Naturally, also the tendency of suspended particles and colloidal particles to enlarge the necessary pressure by clogging the membrane is important. The free sedimentation rate of particles and the flowing properties of suspensions have a great influence on the choice of pump, etc. Naturally, it is very good if the cost of the operation can be analyzed at an early stage.

We can define fixed rules and demands for the osmotic unit.

It must:

– provide proper mechanical support for the membrane,

– provide for uniform distribution of the process solution over the entire membrane surface,

– provide adequate hydrodynamic conditions for the process solution and permeate with minimum loss of energy,

– provide possibilities for minimizing concentration polarization,

– have a high active membrane area/volume ration,

– be easy to dismantle and clean out,

– allow for easy membrane replacement and re-assembly,

– be highly reliable and operationally safe, since it operates under pressure,

– be as free as possible from leakages that could result from pressure-induced changes in dimensional tolerances and handling of the membrane in manufacturing and purification,

– be inexpensive to manufacture, repair and maintain.

According to these rules four types have been developed, all of them functioning for many years:

(1) Tubular units having a small area/volume ration but very good for suspensions.

(2) Spiral-wound units with higher area/volume ration but more sensitive for suspensions.

(3) Plate-and-frame type, first to be used, but is said to be expensive and to have low area/volume ration.

(4) Hollow fibre units in cellulose acetate, nylon and other polyamides have an extremely high area/volume ratio but no possiblities for handling suspensions.

Of course this is a very short presentation and a lot of work has to be done on different types of material in membranes and different ways to support membranes. The intention is to get a membrane suitable for the principle giving a high degree of purification, and high capacity, providing solutions for a long time (years). From the types mentioned other solutions of these types have been developed for a membrane supported by plastic-bound sand.

The way to analyze the situation further must be to concentrate on operation principles and not so much on different examples of what might be done with the principle.

4. The Unit Operation Principle

The well-known unit operation principle means registration of all operations according to the importance of transfer of momentum, heat and mass in the individual operation.

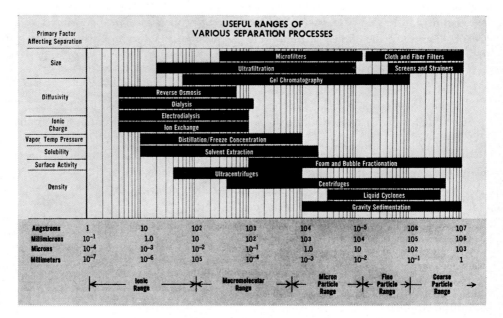

Fig. 1.   Separation processes. (According to Porter and Michaels, 1971)

In this way we get a first group where only transfer of momentum is of real importance, starting with the handling of solid phases in air or other gases (sometimes with the help of small quantities of water), with comminution, sewing, mixing, dust-air cleaning, mechanical aspects of fluidization as examples, and followed by liquid fluidization, separation of solid phases from liquids with or without filters and at last different forms of pressing, extruding and general agglomerations.

The second group contains operations where heat and/or mass transfer defines the principle and naturally the influence of transfer of momentum is taken into consideration. In this group we have several operations on the mass transfer side with and without different types of membranes as dialysis, reverse osmosis, a.o.

We may also get a third group, if we take chemical reaction engineering in consideration and talk of reactor types for chemical or biological transformation.

In today's literature and also in descriptions from manufacturers there is mention of unit operations together with reverse osmosis a.o., but no discription of the place of this peculiar unit operation in relation to other operations in the second group, filtration principles from the first group and in some applications with chemical or biological reactions or transformations.

In the first issue of the new journal *Chemical Technology* (January 1971) Figure 1 was found, a good survey is given of different methods with a primary factor affecting separation taken against largeness of particles. The name for reverse osmosis in several articles is hyperfiltration, which name tells us in a better way about the association with other unit operations, General filtration with different types of filters with or without beds of filtered substances from the liquid or dosated as help substance, which together with hyperfiltration makes it possible to separate even complicated mixtures. Centrifugation and even ultracentrifugation may be used if density differences and the charge of particles to accelerate the separation of them from surrounding liquid are large enough.

The degree of influence from diffusion is different in different operations.

5. Chemical and Biological Engineering

The Institution of Chemical Engineers in London has recently adopted the following definition of Chemical Engineering:

Chemical Engineering is the branch of engineering which is concerned with processes in which materials undergo a required change of composition, energy content, or physical state; with the means of processing; with the resultant products; and with their applications to useful ends. Chemical engineering has its foundations in chemistry, physics and mathematics; its operations are developed from knowledge provided by these disciplines, by the other branches of engineering, and by the applied, biological and social sciences.

The practice of chemical engineering is concerned with: the conception, development, design, improvement and application of processes and their products, the economic development, design, construction, operation, control and management of plants for these processes; and with research and education in these fields.

This definition means that processes involving biochemistry or microbiology are as much the concern of chemical engineers as those based on physical or organic chemistry or metallurgy.

In the opinion of the author this definition does not sufficiently serve the division of units into unit operations, the study of isolated operations and the process engineering aspects.

Biochemical engineering is defined by professor Aiba *et al.* in the following way:

Biochemical engineering is that activity concerned with economic processing of materials of biological character in origin to serve useful purposes. The function of the biochemical engineer is that of translating the knowledge of the micro-biologist and the biochemist into a practical operation. To do this the biochemical engineer must not only be well grounded in the basic engineering principles, but he also must have an appreciation of the biological sciences.

This comment is valid for this definition as well as for the definition of chemical engineering and the possibility for a chemical engineer to work in the whole field of chemical and biological engineering, especially on the unit operation and process engineering side, is very large.

6. Process Engineering

The dividing of unit processes in unit operations has already been mentioned.

The study of these unit operations based on the transfer of momentum, heat, mass and for reactors chemical or biological reaction engineering, follows and provides basic knowledge for the choice of buyable or the construction of new apparatus. Naturally, the special process, the function, and the capacity of apparatus are decisive factors in the choice between alternative unit operations.

The process engineering aspect then is the principle of placing these unit operations again in sequence with input and output fixed to give continuous manufacturing with a minimum of storing and a minimum of risk of a complete stoppage of manufacturing or handling for a long period of time. In this respect it is possible to discuss a smaller or higher number of unit operation steps in order to estimate the cost of different alternatives.

If for example reverse osmosis creates possibilities of diminishing the number of steps in a process, this will make it possible to pay more for this step, and there are also possibilities of getting other operations to work with greater efficiency in more concentrated solutions, and finally environmental problems may be solved or a higher capacity might be reached. It is also possible that a new step in the process such as reverse osmosis may diminish the dimensions of apparatus in further steps and may give values of the type already mentioned.

I will show this by way of some examples. The first one is shown in the Figures 2 and 3 and involves waste water purification.

With waste water quantities taken from different places in two purification works it could be shown (Lindner and Mårtensson, 1971a, b) that the same result was reached independent of where in the sequence the sample was taken. The product water was according to Swedish state regulations approved as raw material for what is called higher degree purification of drinking water at such works.

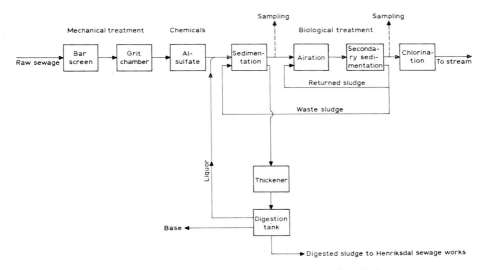

Fig. 2. Flow diagram of Eolshäll sewage works, Stockholm.

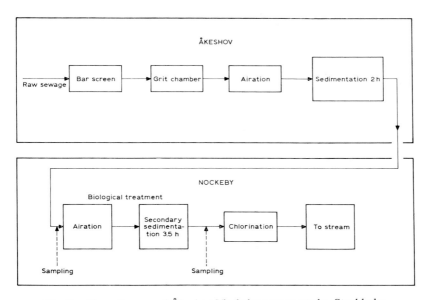

Fig. 3. Flow diagram of Åkeshov-Nockeby sewage works, Stockholm.

If it is possible to diminish the number of steps in waste water purification, natu-rally the cost of reverse osmosis, even if it is high, may give a low total cost.

May I take another example from large laundries in Sweden with 10 to 50 tons of clothes to be washed a day and very often situated near small rivers or lakes. With reverse osmosis it may be possible to purify the water output, if the problem of very small fibres ($< 5 \mu$m) could be solved. In our experiments these fibres tend to enlarge the necessary pressure over the membrane even in small quantities. The problem can surely be solved with an ultrafiltration step but perhaps at too high a cost. Here the benefits of reverse osmosis lie in the fact that production may be enlarged if there is a possibility of handling the water. Naturally the purification is also easier if the water to be handled chemically or biologically is more concentrated.

There is a tendency today to building duplicate systems for waste water gathering in large cities. The so-called urban water here is not pure enough to be put into lakes in an unpurified state. Such water should be sampled and taken through a reverse osmosis unit. If water must undergo ordinary purification we have a large problem with low concentrations and large variations in quantities.

In the chemical industry, including the cellulose industry, we handle a lot of solutions with low concentration of particles, ions, etc. The necessary purification of these solutions can possibly be made through reverse osmosis.

In all these examples the use of a unit operation provides the possibility of reducing the total cost, in spite of the fact that the unit operation itself costs a lot of money.

7. Closing Remarks

In this article I have tried to show how a rather new unit operation fits into a sequence not only as an individual operation but more as a part of a series of unit operations to solve several problems and to minimize total cost even if environmental problems are taken into consideration. This unit operation together with other operations is useful in chemical and biological engineering.

References

Lindner, G.: 1971, Dechema Monografien, No. 64, pp. 295–314.
Mårtensson, B.: 1971a, *Proceedings, Membrane Day*, Univ. of Lund.
Mårtensson, B.: 1971b, FOA 1 Report, A1526–77 (35) (in Swedish).
Porter, M. C. and Michaels, A. S.: 1971, *Chem. Technol.* **1** 56–63.
Sourirajan, S.: 1969, *Reverse Osmosis*, Logos Press Ltd.

of the BOD in the original whey. The plant operates with an inlet pressure of 32–38 kg cm^{-2} to the modules and an outlet pressure of 26–32 kg cm^{-2}.

3.6. SKIMMILK. ULTRAFILTRATION AND HYPERFILTRATION

Perhaps the most important use of ultrafiltration and hyperfiltration in the dairy industry in future will be the ultrafiltration of skimmilk for production of cheese without producing whey.

The process is the following: Water, lactose, and salt are removed from the skimmilk by ultrafiltration. Cream, starter, and rennet enzyme are added, and the cheese is formed without whey production. In this way the whey proteins are kept in the cheese, this gives a protein output which is about 16% higher than with the normal cheese production methods.

The method has been tried out by Maubois *et al.* (1971) at I.N.R.A. in Rennes for camembert, fresh cheese, and goat cheese. The first production plant for ymer has been in operation in Denmark since August 1972 (Nielsen *et al.*, 1972).

When using this method, the main problem has been to obtain the necessary lactose removal in order not to have too acid a cheese. This problem has been solved by using a 28 m^2 module with 17 intermediate flanges and a filtration pressure around 5 kg cm^{-2}.

For camembert production a concentration ratio of 1:6 or 1:7 will be necessary in order to keep the capacity reasonably high and not to get problems with the increased viscosity at the end of the concentration, the temperature is increased from 2–4 °C at the start to 50 °C when a concentration ratio of 1:3 has been reached.

3.7. ELECTROPHORETIC PAINT RECOVERY AND WASTE TREATMENT

This process is used today by most car factories in the world, as it gives a fine layer for the further painting of the car bodies. It appeared, however, that the process caused some waste water problems. In the present systems the thinned paint is cleaned by depositing the colloidal paint particles and leading them to another bath, whereas the aqueous phase with paint residues and various electrolytes make up a waste product.

By introducing ultrafiltration to this cleaning process the following advantages can be obtained:

(1) The waste water will be clean enough to be used for rinsing of the bodies.

(2) The paint concentration in the bath can be kept constant by removing more or less water.

(3) The waste water contains only electrolytes which are not damaging to a possible receiver.

Pilot plants for ultrafiltration and hyperfiltration are now in operation with very good results at a German car factory. The capacity is around 700 l filtrate m^{-2} 24 h.

4. Water Desalination

For water desalination we have installed technical plants for the following applications:

4.1. HEMODIALYSIS

Water treatment plants are installed in several hospitals in Denmark and Germany for water supply to artificial kidney centers. The experimental work has been described by Madsen *et al.* (1970). The main results are the following with modules equipped with type 880 membrane:

(1) More than 95% of all inorganics are removed from tap water.
(2) The produced water is free from bacteria and pyrogenes.

Fig. 6.   View of plant at Copenhagen Municipal Hospital.

(3) The capacity is approx. 660 l m^{-2} 24 h^{-1} at 42 kg cm^{-2}.

(4) The 1.8 m^2 module is ideal for water production for home dialysis.

Figure 6 shows a 7 × 5.4 m^2 plant installed at the Copenhagen Municipal Hospital. This plant has been in operation for 19 months with constant capacity and desalination and without any membrane damage.

4.2. BRACKISH WATER DESALINATION

A 28 m^2 plant is installed at a big camping site in Denmark producing drinking water from brackish water. Membrane type 985 is used. The capacity is approximately 50 l m^{-2} h^{-1}.

4.3. BOILER FEED WATER PRODUCTION

Several plants are in operation for boiler feed water production for low pressure boilers. The new 990 membranes make this process economically favourable.

With the old membrane type 880 the running cost for water production was about 0.25 $ m^{-3} produced water including change of membranes once a year. With the new 990 membranes the costs will be reduced to 0.15 $ m^{-3}.

Investment costs including pumps etc. but exclusive buildings are normally approximately 250 $ m^{-2} filtering area for middle size plants (56–560 m^2). For desalination of tap water to boiler feed water the capacity per m^2 membrane is approx. 1400 l per 24 h.

4.4. SEA WATER DESALINATION

Sea water desalination has been the most popular H.F. subject in the U.S.A. Within our company we have, until recently, only used limited efforts to solve the problems of this application.

The main problems are that the membranes produced so far cannot produce drink-

TABLE III

Results obtained by using H.F. and U.F. on waste waters

Type of waste water	Type of membrane	Capacity 1 m^{-2} h^{-1}	Concentration ratio	KMnO$_4$- figure in waste water mg l^{-1}	KMnO$_4$- figure in permeate mg l^{-1}
Whey, sweet	875	20	1:2	53000	300
Whey, acid	875	20	1:2	53000	2000
Heat-treated household	800	15	1:15	3300	1200
waste water, Farrerr-process	875	11	1:6	3300	580
Yeast waste water	800	11	1:10	18000	2760
Yeast waste water	875	9	1:8	15000	320
Heat-treated potato juice	875	20	1:11	22000	522
Chemically treated	875	45	1:15	700	190
municipal waste water	880	30	1:10	250	90

ing water in a single stage hyperfiltration, and that when used on sea water the membranes have a very limited lifetime. The membrane 999 gives, however, very promising results. A pilot plant has now been running for 6 months with this membrane on sea water (3.5% salt content).

After 6 months' operation the capacity is $200 \, 1 \, m^{-2} \, 24 \, h^{-1}$. The permeate has a salt content of less than 200 ppm. This makes drinking water production with H.F. economically advantageous at remote places without any other water supply than sea water, as the drinking water will cost between 1 and 1.2 \$ per m^3, exclusive interest and amortization, but including change of membranes once a year.

I have now given some of the results we have obtained with H.F. and U.F. The process can, however, be used for many other applications, and we have experience within many other fields, e.g. waste waters (some figures are given in Table III), many solutions in the pharmaceutical industry, coffee, orange juice, etc.

References

Loeb, S. and Sourirajan, S.: 1962, *Advan. Chem. Ser.* **38**, 117.
Lowe, E., Durkee, E. L., Merson, R. L., Ijichi, K., and Cimino, S.: 1969, *Food Technol.* **23**.
Madsen, R. F. and Olsen, O. J.: 1969, *Forskning* **5** (in Danish).
Madsen, R. F.: 1970, Dechema-Monographien, Vol. 64.
Madsen, R. F., Nielsen, B., Olsen, O. J., and Raaschou, F.: 1970, *Nepron* **7**, 545–558.
Maubois, J. L. and Mocquot, G.: 1971, *Le Lait*, LI, **508**, 495–533.
Nielsen, I. K., Bundgaard, A. G., Olsen, O. J., and Madsen, R. F.: 1972, *Reverse Osmosis for Milk and Whey*, Process Biochemistry, pp. 17–20.

PART IX

WATER POLLUTION

Chemical and Biological Effects

BIOCHEMISTRY AND TOXICOLOGY OF THE WASTE WATER
IN ITS EFFECT TO THE RECEIVING WATER

H. LIEBMANN

Bayerische Biologische Versuchsanstalt, Munich, Germany

Abstract. The metabolic physiology of water bacteria is briefly described and the influence of toxic substances and mass transfer is discussed. The knowledge about endo-and exo-enzymes is outlined.

The progress of biochemical and toxicological processes in waste water should not only be observed chemically but above all physiologically. A correct judgement of these processes can only be made after some knowledge of the metabolic physiology of water bacteria has been acquired.

Therefore, in practicing water economy the statements on the growth of water bacteria are often not very accurate – because we are inclined to define bacteria in water as generally disadvantageous organisms. We tend to immediately think of pathogenic forms, which are found in domestic sewage.

In waste water and receiving water apathogenic and other mostly useful forms preponderate. These forms play a predominant role in the different methods of sewage purification. In the anaerobic as well as in the aerobic phase of sewage purification and in natural self-purification in receiving water, the predominant micro-organisms are bacteria. An intensification of sewage purification and a partial biogenous detoxification in the waste water is possible if the growth of water bacteria is encouraged. 1 ml of crude sewage can contain more than 1 million germs – of these more than 90% are not disadvantageous but are useful for decomposition of the organic substance; that is, for natural self-purification.

Bacteria are only able to decompose organic substances in waste water if they find optimal biological conditions. Bacteria are very sensitive to changes of temperature, changes in chemical structure and radiation. Most bacteria will die in the temperature-region between $+50\,^{\circ}C$ and $75\,^{\circ}C$. Temperatures of $-20\,^{\circ}C$ cannot be tolerated for a long time, apart from the formation of spores, which are resistant to high and low temperatures. In purification plants or in surface waters bacteria can adapt to temperature and water conditions. Here the principle supposition is that the pH regions lay between 5.5 and 8.5 .Bacteria are capable of becoming acclimatized or of gradually adapting to food which is present in the waste water. Here attention must be paid to the fact that in many industrial wastes the microorganisms find chemical substances which are primarily not present in nature, so that they have to accomplish metabolic conversions without biogenetical preparation (Huber, 1968). Heavy metals, e.g., manganese, cobalt, nickel, chromium, arsenic, cadmium, lead, iron, zinc, tin, gold, silver, mercury, and copper are toxic at certain concentrations. So far there are no standard methods of determining the toxicity of these heavy metals against microorganisms.

G. Lindner and K. Nyberg (eds.), Environmental Engineering, 333–337. All Rights Reserved

Every examiner takes other bases of relations so that the results are not comparable. In the handbook *Frischwasser- und Abwasserbiologie*, Vol. II (Liebmann, 1960), in a chapter which has more than 300 pages, together with Stammer, we have taken the trouble to register the toxicity of different chemical constituents in waste water.

1. Discussion of Toxicity

The toxicity of the same heavy metal differs very much because the methods used by the different experimenters are not standardized. Some used the so-called dilution method (BOD) as a basis for their experiments, that is to say they use industrial waste waters which are artificially diluted, and hereby they start from the idea that the original toxicity is the same in spite of this dilution, an approximation that of course in many cases is not correct. Besides that for the metabolic physiology of water bacteria, it cannot be without importance that the BOD determination methods until now with the aid of the dilution method has precisely the same quantity of oxygen at its disposal which is present in the mixture of waste water and dilution water. Water bacteria, which will be tested for its function of oxygen depletion and the possible influence of heavy metals have no optimal life conditions. The observed degenerative processes distort the results.

Besides the above-mentioned errors, which are compounded by the fact that the measurements are carried out over a short period, another mistake creeps in through arbitrarily measuring the inhibition of bacteria for only a short time, e.g. for 8 to 10 h. Added to these factors there is also the influence of heavy metals, which have effects on the bacteria between 1 and 2 h, others between 24 and 48 h, and finally somewhere the effects can only be seen after 5 days. Inhibition measurements of the growth of water bacteria influenced by bimetals will in future be performed only with the planimetrical method, and every time it must be possible to represent the inhibition area. A fully automatic apparatus such as the Sapromat A6 which is described in some publications is required.

2. Mass Transfer Problems

In the biological step of sewage purification as well as self-purification in receiving water a special problem is the supply of air to the bacteria cell and the removal of metabolic products from this. If metabolic processes of bacteria cell are very intensive – which generally is the case in aerated and mechanically pre-sedimented waste water – the bacteria cell absorbs the organic substance, but at the same time many digested organic substances will be separated. These separations form a so-called physiological circle around the bacteria cell, and it constantly reduces the replacement according to the growth of the cell. To advance the metabolic processes of the waste water bacteria this physiological circle must be destroyed. In nature this destruction will be carried out most efficiently in a fast flowing flat mountain river, which has a high turbulence because of many stones and a large flow velocity. This turbulence almost

BIOLOGICAL NITROGEN REDUCTION STUDIED AS
A GENERAL MICROBIOLOGICAL ENGINEERING PROCESS

BENGT HULTMAN

Dept. of Water and Sewage Technique and Water Chemistry,
The Royal Institute of Technology, Stockholm, Sweden

Abstract. The different factors influencing biological nitrogen removal are briefly described. With the help of general formulae of bacterial growth, oxidation of the energy source and uptake of the electron acceptor, the kinetics of nitrification and denitrification are discussed. The developed formulae have been used to illustrate dimensioning of different process stages of plants for nitrogen removal.

This paper will briefly describe the kinetics of nitrogen removal by microorganisms. The purpose of the paper is to show that dimensioning of plants for removal of nitrogen may be done in a similar way as dimensioning of other biological processes in fermentation industries or in wastewater treatment. The paper is based on studies of biological nitrogen removal, which began at Vattenbyggnadsbyrån, Stockholm, about eight years ago and which were followed up at the institutions of Water Supply and Sewerage and Water Chemistry, The Royal Institute of Technology, Stockholm. The work from these institutions has been reported in six publications (Edholm *et al.*, 1970; Gustafsson *et al.*, 1967, 1968, 1969; Hultman, 1971; and Sven-Nilsson *et al.*, 1966).

There are two ways by which nitrogen can be removed by microorganisms. The first way, assimilative nitrogen removal, is due to the formation of excess sludge containing nitrogen. Due to this mechanism nitrogen reduction of between 20 and 50% is normally obtained in a conventional activated sludge process.

Nitrogen can also be eliminated by dissimilative nitrogen removal or denitrification. In this process the oxygen content in nitrate or nitrite in the absence of air is used by microorganisms in oxidation of organic substances. Nitrate and nitrite are simultaneously reduced to nitrogen gas, N_2.

The nitrogen in municipal wastewater mainly consists of ammonia and urea. Before denitrification can take place ammonia and urea must be oxidized to nitrite and nitrate by special microorganisms. This process is called nitrification.

1. Kinetics of Growth of Microorganisms

The growth of different bacteria follows very similar formulae. In order to grow chemical energy must be available in a usable form as in energy-rich phosphate groups. These phosphate groups are formed in a series of oxidation-reduction reactions within the bacteria. The oxidation-reduction may be described as a dehydrogenation of an energy source followed by transfer of hydrogen or electrons to an ultimate acceptor. In Table I energy sources and electronacceptors are exemplified for different bacteria.

G. Lindner and K. Nyberg (eds.), Environmental Engineering, 339–350. All Rights Reserved
Copyright © 1973 by D. Reidel Publishing Company, Dordrecht-Holland

TABLE I

Examples of energy sources and electronacceptors for some bacteria

Energy source	Electronacceptor	Type of bacteria
Organic material	O_2	Heterotrophic aerobic bacteria
NH_3, NO_2^-	O_2	Nitrification bacteria
H_2	O_2	Hydrogemonas
S, H_2S, $S_2O_3^{2-}$	O_2	Thiobacillus
Fe^{2+}	O_2	Iron bacteria
Organic material	NO_2^-, NO_3^-	Denitrification bacteria
Organic material	SO_4^{2-}	Desulfovibrio
Organic material	CO_2	Heterotrophic anaerobic bacteria
H_2	CO_2	Methanobacterium

TABLE II

Rate equations for bacterial growth

Bacterial growth:

$$V_{X\alpha} = \frac{dX\alpha/dt}{X\alpha} = \mu - b - v\,[h^{-1}]. \tag{1}$$

Uptake rate of the energy source:

$$V_s = \frac{ds/dt}{X\alpha} = -\frac{\mu}{Y_s}\left[\frac{\text{g energy source}}{\text{g} \quad \text{VSS}\cdot\text{h}}\right]. \tag{2}$$

Uptake rate of electrons from the energy source:

$$V_{es} = \frac{ds_{es}/dt}{X\alpha} = -\frac{\mu}{Y_{es}}\left[\frac{\text{electron eqv.}}{\text{g VSS}\cdot\text{h}}\right]. \tag{3}$$

Uptake rate of oxygen:

$$V_{O_2} = \frac{dc_{O_2}/dt}{X\alpha} = -\frac{\mu}{Y_{O_2}} - b_{O_2}\left[\frac{\text{g oxygen}}{\text{g VSS}\cdot\text{h}}\right]. \tag{4}$$

Uptake rate of nitrate:

$$V_{NO_3^-} = \frac{dc_{NO_3^-}/dt}{X\alpha} = -\frac{\mu}{Y_{NO_3^-}} - b_{NO_3^-}\left[\frac{\text{g } NO_3^- - N}{\text{g VSS}\cdot\text{h}}\right]. \tag{5}$$

Uptake rate of electrons by the electron acceptor:

$$V_{ee} = \frac{dc_{ee}/dt}{X\alpha} = -\frac{\mu}{Y_{ee}} - b_{ee}\left[\frac{\text{electron eqv.}}{\text{g VSS}\cdot\text{h}}\right]. \tag{6}$$

Where μ = growth rate function $[h^{-1}]$, X = sludge concentration $[g\ VSS\ m^{-3}]$, α = viable part of the sludge, t = time $[h]$, b = rate of endogeneous respiration, v = rate of bacterial death $[h^{-1}]$, V = modified growth rate function, Y = yield coefficient, s = concentration of the energy source, c = concentration of the electron acceptor.

TABLE III

Equations for the growth rate function (μ), apparent sludge yield (Y_{app}), and the viable part (α) of the sludge (Hultman 1971)

The growth rate function
Influence of the energy source:

$$\mu(s) = \frac{\mu_{max}s}{K_s + s}. \tag{7}$$

Influence of the electron acceptor:

$$\mu(c) = \frac{\mu_{max}c}{K_c + c}. \tag{8}$$

Temperature influence:

$$\mu_{max}(T) = \mu_{max}(20\,^\circ C)\,10_1^{k(T-20)}. \tag{9}$$

$$K_s(T) = K_s(20\,^\circ C)\,10_2^{k(T-20)}. \tag{10}$$

$$K_c(T) = K_c(20\,^\circ C)\,10_3^{k(T-20)}. \tag{11}$$

pH influence:

$$\mu_{max}(pH) = \frac{\mu_{max}(pH)_{opt}}{1 + k_4[10^{/pH}opt^{-pH/} - 1]}. \tag{12}$$

$$K_s(pH) = K_s. \tag{13}$$

$$K_c(pH) = K_c. \tag{14}$$

From formulae (7)–(14):

$$\mu(s, c, T, pH) = \frac{\mu_{max}(pH_{opt}, 20\,^\circ C)\,sc10_1^{k(T-20)}}{[K_s(20\,^\circ C)\,10_2^{k(T-20)} + s]\,[K_c(20\,^\circ C)\,10_3^{k(T-20)} + c]}$$

$$\times \frac{1}{[1 + k_4(10^{/pH}opt^{-pH/} - 1)]}. \tag{15}$$

Apparent yield

$$Y_{app} = Y_s\left[1 - \frac{b}{b + v + 1/G}\right]. \tag{16}$$

Viable part of the sludge

$$\alpha = \frac{k_5}{1 + vG} \tag{17}$$

where μ_{max} = maximum growth rate, T = temperature $^\circ C$, pH = pH-value, pH_{opt} = optimal pH-value, K_s, K_c, k_1, k_2, k_3, k_4 and k_5 = const, G = sludge age defined by Formula (18).

$$G = \frac{1}{\mu - b - v}. \tag{18}$$

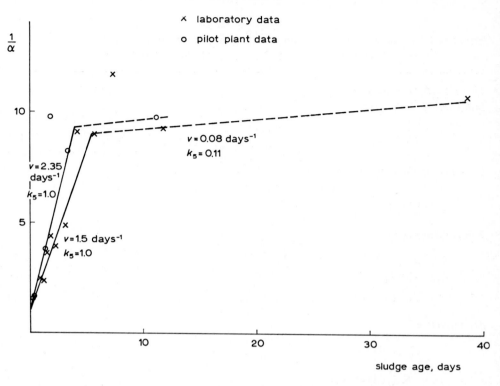

Fig. 1.   Steady state viability plotted with the help of Formula (17). (Data according to Weddle and Jenkins (1971).)

Formulae relating the growth of bacteria, the uptake rate of the energy source and the electronacceptor are shown in Table II. The growth rate function μ in these equaitons is a function of the temperature, pH-value, limiting substrate concentration, toxic substances and the mixing conditions. Formulae describing the influence of temperature, pH-value and limiting substrate concentration are summarized in Table III.

The pH-dependence is based on a model in which a deviation of the pH-value from the optimal reduces the activity of the bacteria according to the mechanism of non-competitive inhibition. This means that the value of K_s in Formula (7) is not influenced by the pH-value (Hultman, 1971).

Depending on the endogeneous respiration and the death of bacteria the viability of the sludge and the apparent yield (sludge produced/energy source removed) diminish if the sludge age increases (see Formulae (16) and (17) in Table III). The formula relating the viability of the sludge is based on models described by McCarty and Brodersen (1962), and Sinclair and Topi wala (1970). The use of Formula (17) for data by Weddle and Jenkins (1971) for the activated sludge process is illustrated in Figure 1.

2. Rate Constants for Bacterial Growth

Although the growth of different bacteria may be described by similar formulae the different energy sources and electron acceptors will give very varying values of the rate constants. The maximum growth rate and the yield coefficients may approximately be calculated by use of thermodynamical data (Hultman, 1972; McCarty, 1971).

It may be shown (Hultman, 1972) that the maximum uptake rate of organic substances is approximately the same, if oxygen is supplied from air, nitrate or sulphate. This means that it is possible to use data from aerobic treatment of sewage for dimensioning of a denitrification stage. The rate of growth of nitrification bacteria is very low depending on low amount and bacterial efficiency of the energy obtained in oxidation of ammonium and nitrite.

3. Assimilative Nitrogen Removal

In assimilative nitrogen removal different nitrogen sources (ammonium, nitrite and nitrate) are used for synthesis of protoplasm of the microorganisms. The assimilative nitrogen reduction β is approximately (Hultman, 1971)

$$\beta = \frac{[Y_{app}(s_0 - s) - X_{out}]\gamma}{z_0}, \tag{19}$$

where s_0 = incoming organic substrate concentration, s = organic substrate concentration in the effluent, X_{out} = sludge concentration in the effluent, γ = nitrogen part in the sludge, z_0 = incoming nitrogen concentration.

In order to illustrate Formula (19) the following process conditions are assumed:

$$Y_{app} = 0.65 \text{ g VSS g}^{-1} \text{ BOD}_5 \text{ removed},$$
$$s_0 - s = 200 \text{ g BOD}_5 \text{ m}^{-3},$$
$$X_{out} = 10 \text{ g VSS m}^{-3},$$
$$\gamma = 0.1 \text{ g N g}^{-1} \text{ VSS},$$
$$z_0 = 30 \text{ g N m}^{-3}.$$

If these values are inserted in Formula (19) an assimilative nitrogen reduction of 40% is obtained.

4. Nitrification

Nitrification is mainly caused by two kind of autotrophic bacteria, Nitrosomonas and Nitrobacter. Nitrosomonas oxidize ammonium to nitrite according to the reaction:

$$NH_4^+ + 1.5O_2 \rightarrow NO_2^- + H_2O + 2H^+,$$

and Nitrobacter oxidize nitrite to nitrate according to the reaction:

$$NO_2^- + 0.5O_2 \rightarrow NO_3^-.$$

Thus the total reaction in oxidizing ammonium to nitrate is:

$$NH_4^+ + 2O_2 \rightarrow NO_3^- + 2H^+ + H_2O .$$

Present knowledge of the kinetics of nitrification is essentially based on works by Downing *et al.* (1964), who have used Monod's or Michaelis–Menten's formula (i.e. Formula (7) in Table III) for describing nitrification. The maximum growth rate of nitrification bacteria is much lower than that of heterotrophic bacteria. Any type of microorganism in a continuous activated sludge process is washed out if the amount of the microorganism in question produced in the reactor is less than the amount removed as excess sludge. Nitrification bacteria are washed out if:

$$G < \frac{s_0 + K_s}{\mu_{max} s_0} \approx \frac{1}{\mu_{max}}, \tag{20}$$

where G = sludge age, s_0 = influent ammonium concentration.

The value of K_s for nitrification bacteria is low and may as a rough approximation be neglected in comparison with s_0. Because of the low growth rate of nitrification bacteria a high sludge age is necessary if nitrification is to be obtained. In a continuous activated sludge process the sludge age of nitrification bacteria and heterotrophic bacteria may be assumed to be equal.

The value of μ_{max} for the nitrification bacteria is dependent on the values of the temperature and pH according to Formulae (9)–(14) in Table III. The use of Formula

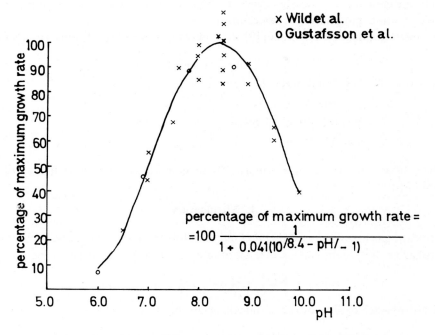

Fig. 2.   Growth rate of Nitrosomonas as a function of the pH-value.

(12) is illustrated in Figure 2 with data from Gustafsson *et al.* (1969) and Wild *et al.* (1971). In Table IV normal values of rate constants in nitrification in an activated sludge process are shown.

TABLE IV

Rate constants in nitrification

Constant	Value
$\mu_{max}(20°C, pH_{opt})$	0.3–0.5 days^{-1}
K_s	0.5–2.0 g N m^{-3}
Y	≈ 0.05 g VSS g^{-1} N
pH$_{opt}$	8.0–8.4
k_1 (cf. Formula (9))	0.028–0.038 °C^{-1}
k_4 (cf. Formula (12))	≈ 0.04

Under stationary conditions and complete mixing the percentage nitrification $\eta_s = 100(s_0-s)/s_0$ may be written

$$\eta_s = 100\left[1 - \frac{K_s}{s_0(\mu_{max}(T, pH) G - 1)}\right] \tag{21}$$

In order to illustrate Formula (21), η_s has been drawn as a function of G in Figure 3. The following constant values have been chosen: $s_0 = 30$ g NH$_3$–N m^{-3}, $\mu_{max}(T, pH) = = 0.12$ days^{-1} (corresponding to nitrification at about 8°C), $K_s = 0.5$ g NH$_3$–N m^{-3} and 2.0 g NH$_3$–N m^{-3}, respectively.

As shown in Figure 3 no nitrification is obtained for a sludge age below about 8.5 days. To secure a high degree of nitrification at about 8°C a sludge age of around 15 days is necessary.

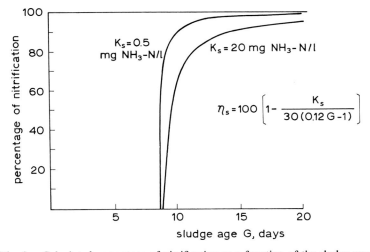

Fig. 3.   Calculated percentage of nitrification as a function of the sludge age.
(According to Gustafsson *et al.*, 1969.)

5. Denitrification

Denitrification may be regarded as a two-step process, where nitrate is reduced via nitrite to nitrogen gas. This is illustrated by the following reactions with methanol as a carbon source:

$$NO_3^- + \tfrac{1}{3}CH_3OH \rightarrow NO_2^- + \tfrac{1}{3}CO_2 + \tfrac{2}{3}H_2O \tag{a}$$

$$NO_2^- + \tfrac{1}{2}CH_3OH \rightarrow \tfrac{1}{2}N_2 + \tfrac{1}{2}CO_2 + \tfrac{1}{2}H_2O + OH^- . \tag{b}$$

$$NO_3^- + \tfrac{5}{6}CH_3OH \rightarrow \tfrac{1}{2}N_2 + \tfrac{5}{6}CO_2 + \tfrac{7}{6}H_2O + OH^- \tag{c}$$

Aerobic oxidation of methanol may be written:

$$CH_3OH + \tfrac{3}{2}O_2 \rightarrow 2CO_2 + 4H_2O \tag{d}$$

With the help of the reaction Formulae (a)–(d) the corresponding amount of oxygen supplied in denitrification is obtained:

$$NO_3^- \rightarrow \tfrac{1}{2}O_2 + NO_2^- \tag{e}$$

$$NO_2^- + \tfrac{1}{2}H_2O \rightarrow \tfrac{3}{4}O_2 + \tfrac{1}{2}N_2 + OH^- . \tag{f}$$

$$NO_3^- + \tfrac{1}{2}H_2O \rightarrow \tfrac{5}{4}O_2 + \tfrac{1}{2}N_2 + OH^- \tag{g}$$

For every g NO_3^-–N reduced to nitrogen gas 2.86 g oxygen is supplied according to Formula (g).

Denitrification occurs simultaneously with the oxidation of organic substances. In the oxidation of 1 g COD Y_s g cellsubstance is formed calculated as COD. The difference $(1 - Y_s)$ corresponds to oxygen consumption in the oxidation of 1 g COD. The value of Y_s is about 70% of the value of Y_s under aerobic conditions and is about 0.4 g COD/g COD (Edholm et al., 1970). If the influent nitrogen in an activated sludge plant consists of nitrate–nitrogen the removed nitrate E_N depends both on assimilation and denitrification according to the formula:

$$E_N = QY_{app}(s_0 - s)\gamma + \frac{Q(s_0 - s)(1 - Y_{app})}{2.86}, \tag{22}$$

where Q = flow rate; γ = nitrogen part in the sludge.

The ratio A of necessary COD and nitrate–nitrogen removed may be written:

$$A = \frac{Q(s_0 - s)}{Q(s_0 - s)(Y_{app}\gamma + (1 - Y_{app})/2.86)} = \frac{2.86}{2.86 Y_{app}\gamma + 1 - Y_{app}}. \tag{23}$$

If $\gamma = 0.14$ g N g^{-1} COD and Y_{app} equal to 0.1 and 0.4 the value of A is 3.0 and 3.8, respectively. By use of methanol with a chemical oxygen demand of 1.5 g COD g^{-1} methanol as a carbon source a ratio of 2.0 and 2.5 g methanol g^{-1} NO_3^-–N, respectively, is obtained. The calculations may easily be extended to include the influence of oxygen and nitrite (Hultman, 1971).

Oxidation of ferrous ions is also possible with dissolved oxygen in the alkaline pH-region. This method has been tested both in Västerås and Stockholm (Nockeby) on pilot-plant scale. By aeration in the first flocculation tank instead of agitation, the reaction rate at neutral pH should occur more rapidly. The oxidation rate is very slow in the acidic range and markedly dependent on the pH-value as well as on the oxygen pressure. The results clearly indicated, particularly regarding the flocculation and sedimentation, that the oxidation should be performed in the alkaline pH-region after the addition of lime.

According to Krause et $al.$ (1937), calcium ferrite may be precipitated in the presence of lime. At Nockeby 200 mg l^{-1} lime besides 125 mg l^{-1} ferrous sulphate was required to achieve a phosphorus reduction to about 0.5 mg P l^{-1}.

The experiments were continued at Orsa sewage treatment works on a technical scale. The best purification effect by post-precipitation of water with a high oxygen concentration was obtained with 150–200 g m^{-3} ferrous sulphate and 100–150 g m^{-3} lime (pH 9–9.5). In these cases the BOD$_7$- and phosphorus-removal in the chemical treatment were about 80 and 90%, respectively (remaining BOD$_7$ and total phosphorus = about 20 mg O$_2$ l^{-1} and 0.5–1.0 mg P l^{-1}). The residual amount of iron after sedimentation was about 3 mg l^{-1}.

At Borlänge the first chemical plant with chlorinated ferrous sulphate is under construction.

Thickening tests in long-tubes have shown that alum sludge from chemical post-treatment may be concentrated to 1.5–3.5% dry matter, ferric sludge to 1–3.5% and lime sludge to about 10%. The varying results have been obtained from different sewage plants with different composition of the sewage before precipitation. The results emphasize, that more attention should be paid to the influence of the rheological parameters on the thickening properties of the sludge.

In Sweden centrifuges or pressure sieves have been used to dewater chemical sludge. Addition of polymers is required. The final disposal of 2–3 million m^3 sludge per year in Sweden is at present a major problem. When introducing chemical post-treatment, the increase of sludge in a biological treatment plant with digestion, based on dry matter, will be about 40% with alum, 50% with ferric salts and 250–500% with lime without recovery. Lime recovery by thermal decomposition of CaCO$_3$ is a feasible process in large plants. Research on aluminium recovery is proceeding (Ericsson and Lundberg, 1970), but this is more difficult than lime recovery because phosphorus in solution after treatment of the sludge with acid or alkali has to be separated from dissolved aluminium.

3.3. TEST OPERATION IN PLANT SCALE

At present the authorities in Sweden seem to favour chemical post-treatment after biological treatment. Interest in simultaneous chemical and biological treatment has diminished considerably after the successful operation of chemical pretreatment with alum before biological treatment in Stockholm. There is, however, a plant at Hällefors in operation with simultaneous chemical and biological treatment. The results with

$90 \ g \ m^{-3}$ alum (8% Al) have been better than expected (80–85% phosphorus reduction and about 90% BOD_5 reduction). Test operation with simultaneous chemical and biological treatment was performed already in 1964 at Botkyrka sewage plant (Ericsson, 1967). The chemical dose was only 28 mg l^{-1} $FeCl_3 \cdot 6H_2O$. The total phosphorus content in the effluent was reduced from 4.5 to 2.2 mg $P \ l^{-1}$, and the soluble phos-phosphorus from 4.1 to 1.1 mg $P \ l^{-1}$ on the average. Recently test operation with the addition of ferrous sulphate in the inlet of the aeration tanks started in Sweden at some smaller sewage plants (e.g. Vikmanshyttan and Säter: 2000–7000 p). The anal-yses indicate that oxidation from ferrous to ferric state occurs in the aeration tanks. During a trial period of 1–2 months the iron content in the effluent was below 1 mg $Fe \ l^{-1}$. The content of total phosphorus in the effluent was reduced from about 2 mg $P \ l^{-1}$ to 0.2–0.5 mg $P \ l^{-1}$. Any deterioration of the activated sludge process was not observed; on the contrary the sedimentation in the clarification tanks was im-proved.

Test operation with pre-precipitation on a technical scale has been performed by Vattenbyggnadsbyrån with the addition of ferric chloride and alum at Botkyrka and Jönköping, respectively (Ericsson, 1972). Using pre-precipitation, three parameters are of great significance to the activated sludge process. These are pH, BOD and phosphorus content.

The lowest BOD-content in pre-settled sewage for steady state may be calculated according to the following balance equation, if the content of suspended inorganic and inert organic solids in the pre-settled sewage is neglected and complete mixing is assumed:

$$\frac{X\theta}{G} = Y_{app}(S_0 - S) = X_{out}.$$

where X = suspended solids content in aeration tanks, θ = residence time, G = sludge age, Y_{app} = apparent sludge yield, S_0 = substrate concentration in pre-sedimented sew-age, S = substrate concentration in aeration tanks and effluent, X_{out} = suspended solids in effluent.

It is evident from calculation examples with the formula above that the content of suspended solids in the aeration tanks can vary greatly, depending on the operation conditions of the activated sludge process.

The organic substrate generally limits the growth rate of the activated sludge since nutrients and oxygen are supplied in excess. With pre-precipitation, however, phos-phorus may become growth-rate limiting for the activated sludge in the case of too high phosphorus reduction in the pre-treatment stage. The common value of the phos-phorus content in activated sludge varies between 1 and 3%. Soluble phosphorus, which is easier to assimilate by micro-organisms than suspended phosphorus, is more important than total phosphorus.

The trial operation with pre-precipitation at Botkyrka sewage plant during the peri-od 8 June–24 June 1971 was short, and definite conclusions may not be drawn. The results of the operation were, however, generally good and indicate that the total